# Threads and Spinons Theory:

## Redefining Gravity, Light, Cosmos, and Quantum Mechanics Without Paradoxes or Extra Dimensions

A Unified, Testable Framework for the Universe—
Rooted in Logic, Not Abstractions.

by

Samir Hanna safar

# Copyright Statement

## © 2025 Samir Hanna Safar. All Rights Reserved.

## All rights reserved.

Threads and Spinons Theory
Samir Hanna Safar

## Dedication

To my wife, Lori,

The one who believed in me when no one else did.

The one who saw my creativity and imagination not as foolish dreams but as the seeds of something greater. The one who lifted me through every failure, who reminded me to keep going and never give up on my inventions, ideas, and this theory. This book exists because of your unwavering support, encouragement, and faith in me.

Thank you for always standing by my side, believing in my vision, and pushing me forward when the world told me to stop.

This is for you.

With all my love and gratitude,

Samir Hanna Safar

Threads and Spinons Theory
Samir Hanna Safar

Welcome to the

# Threads and Spinons Light Theory.

Threads and Spinons Theory
Samir Hanna Safar

Threads and Spinons Theory
Samir Hanna Safar

# Table of Contents

## Preface

**Why This Book?** ….. 1

**Introduction:** A Journey Through Physics, Rejection, and AI
…... 3

## Part I: The Origins of the Theory

**Chapter 1: More Historical Context** - The Evolution of
Physics and Where It Went Wrong ...... 11

**Chapter 2: Connections to Existing Physics** - How This
Theory Corrects and Expands on General Relativity, Quantum
Mechanics, and Electromagnetism .... 17

**Chapter 3: Addressing Objections** - Why Mainstream
Physics Falls Short and How This Theory Fills the Gaps .... 27

## Part II: The Core Scientific Framework

**Chapter 4: Light – A Thread- Based Phenomenon** -
Redefining Light: A Continuous Thread Model ...... 37

**Chapter 5: The Light Spectrum** - How Spinon Stacks
Define Electromagnetic Waves ...... 45

**Chapter 6: The Six Fundamental Laws of the Threads and Spinons Theory** ...... 53

**Chapter 7: The 20 Principles of the Threads and Spinons Theory** ...... 59

**Chapter 8: The Spinon Energy Unit (SEU)** - A New Measure of Energy ...... 67

**Chapter 9: Replacing Einstein's Theory with SEU** - A New Framework for Energy, Gravity, and Motion ..... 73

**Chapter 10: The Electron** - A Thread- Based Structure in the Threads and Spinons Theory ..... 81

**Chapter 11: The Proton -** A Structured Thread Formation in the Threads and Spinons Theory ...... 89

**Chapter 12: The Neutron -** A Transitional Thread Structure in the Threads and Spinons Theory ...... 97

**Chapter 13: The Nuclear Bond** - A Thread- Based Interaction in the Threads and Spinons Theory ...... 107

**Chapter 14: The Electron Bond** - A Thread- Based Mechanism for Chemical Bonding in the Threads and Spinons Theory. ..... 115

**Chapter 15: The Electromagnetic Field** - A Thread- Based Interaction in the Threads and Spinons Theory ..... 123

**Chapter 16: The Nature of *Electrical Current* in the Threads and Spinons Theory** ...... 131

**Chapter 17: The Nature of the *Magnetic Field* in the Threads and Spinons Theory** ...... 139

**Chapter 18: The Nuclear Cocoon** - A Thread- Based Shield in the Threads and Spinons Theory ...... 149

**Chapter 19: The Outer Atomic Surface** – A Thread- Based Model of Atomic Boundaries in the Threads and Spinons Theory ...... 159

**Chapter 20: Energy Transfer and Heat Transfer in the Threads and Spinons Theory** ...... 169

# Part III: Experiments and Quantum Phenomena

**Chapter 21: The Double- Slit Experiment** - A Thread- Based Explanation Without Paradoxes ...... 181

**Chapter 22: Testing the Double** - Slit Experiment with Electrons and Protons ...... 189

**Chapter 23: The Schrödinger's Cat** Paradox - A Thread- Based Rebuttal to the Copenhagen Interpretation ...... 197

**Chapter 23-B: Quantum Entanglement** - A Thread- Based Explanation for "Spooky Action at a Distance" ...... 205

# Part IV: Cosmology and the Structure of the Universe

**Chapter 24: Gravity as Thread Tension** - A Physical Replacement for Spacetime Curvature ...... 215

**Chapter 25: The Rejection of the Big Bang** - A Universe That Grows, Not Explodes ...... 223

**Chapter 26: The Expansion of the Universe** - Why There Is No Edge or Center .... 231

**Chapter 27: Rethinking Black Holes** - Ultra- Dense Thread Cores, Not Singularities ...... 239

## Part V: Unifying All Forces and Time

**Chapter 28: Experimental Tests for the Universal Thread Network** ..... 251

**Chapter 29: Technological Advancements with Threads and Spinons Theory** ...... 261

**Chapter 30: The Nature of *Consciousness*** - A Thread-Based Explanation of Thought and Awareness ...... 269

**Chapter 31: Quantum Memory** - Information Storage and Retrieval in the Threads and Spinons **Theory** ...... 277

**Chapter 32: The Expansion of the Universe** - Why There Is No Edge or Center ...... 285

**Chapter 33: Rethinking Dark Matter** - The Threads and Spinons Explanation ...... 293

**Chapter 34: The Problem with** $c^2$ Why Einstein's Equation is Arbitrary 301

**Chapter 35: The Formation of Atomic Nuclei** - A Threads and Spinons Perspective ...... 307

**Chapter 36: Quantum Computing and Subatomic Memory Storage in the Threads and Spinons Framework** ...... 315

**Chapter 37: Why We Can See Deep Into Space** - The Spinon Transport Mechanism ...... 323

**Chapter 38: Unifying All Forces** - Gravity, Electromagnetism, and Nuclear Interactions Under the Threads and Spinons Theory ...... 331

**Chapter 39: The Origin and Evolution of the Universe Without the Big Bang** ...... 339

**Chapter 40: The Nature of Time - Eliminating Time as a Fourth Dimension** ...... 347

**Chapter 41: The Rewriting of Black Holes - The Core Concept** ...... 355

**Chapter 42; Experimental Proposals for Testing the Threads and Spinons Theory** ...... 363

# VI: Closing Chapter

**Closing Chapter: A Theory in Progress - The Path to Refinement and Justification** ...... 373

**References** ...... 379

Threads and Spinons Theory
Samir Hanna Safar

**Why This Book?**

A physicist does not write this book in a university office. It is not filled with dense equations that only a handful of experts can decipher. Instead, it is written by someone who sees physics differently—someone who believes in simplifying the most complex mysteries of the universe.

I am not here to replace physics. I am here to offer an alternative way to understand it.

If you have ever questioned the abstract nature of modern physics, wondered if there was a more intuitive explanation for quantum mechanics, or felt we were missing something fundamental, this book is for you.

This is my journey.

Moreover, this is the **Threads and Spinons Light Theory**.

Threads and Spinons Theory
Samir Hanna Safar

## Introduction

## A Journey Through Physics, Rejection, and AI

## My Journey – From an Unlikely Student to a Theoretical Thinker

In the early 1990s, years after finishing college, I began sketching an idea that had been on my mind for some time. How physics was taught—especially concepts like hydrogen bonding, the fourth dimension, and spacetime—felt unnecessarily abstract. While others accepted these ideas without question, I felt something was missing, like physics had drifted away from physical reality.

Threads and Spinons Theory
Samir Hanna Safar

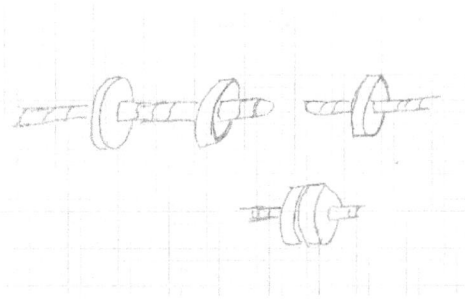

I am not a physicist or mathematician—just someone who took college courses in these subjects. I barely passed them, scraping by with C-minus grades. However, that never stopped my mind from questioning the foundations of modern science. I believed there had to be a more intuitive way to explain the universe that did not rely on complex equations or unprovable dimensions.

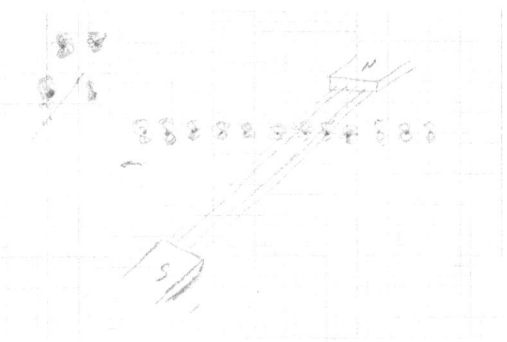

Determined to bring my ideas to life, I started sketching my concepts and commissioned people to create short videos visualizing them. I envisioned a model where point-like particles did not govern the universe but by continuous threads and spinning elements—structures that dictated motion, energy, and form. However, my ideas remained

unpublished without the mathematical expertise or
institutional credentials.

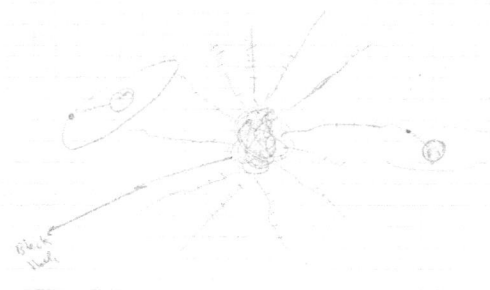

**The Breakthrough:** AI and the Power of Technology
For years, I faced rejection. Scientific journals dismissed my
work—not because of its content but because I lacked
affiliation with a prestigious institution. The gatekeeping in
academia felt impenetrable. Then, artificial intelligence
arrived.

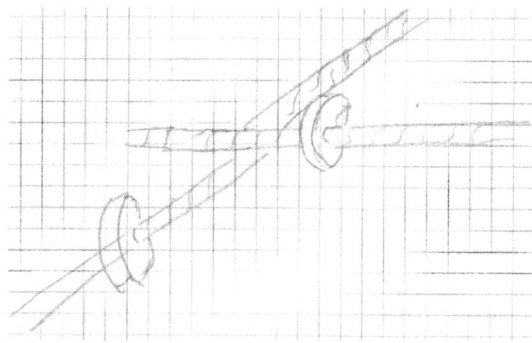

AI changed everything. It allowed me to refine my ideas,
analyze complex relationships, and structure my Theory in a
way I never could. I could finally present my work in a
tangible, organized format with AI. It became an equalizer,
enabling me to explore physics outside traditional academia.

Threads and Spinons Theory
Samir Hanna Safar

Still, skepticism followed. I encountered criticism even after sharing my Theory on YouTube and through short videos. However, something else happened—I gained traction. Some dismissed my ideas outright, but others were intrigued. Over time, the Theory evolved into something more complete, something worthy of a book.

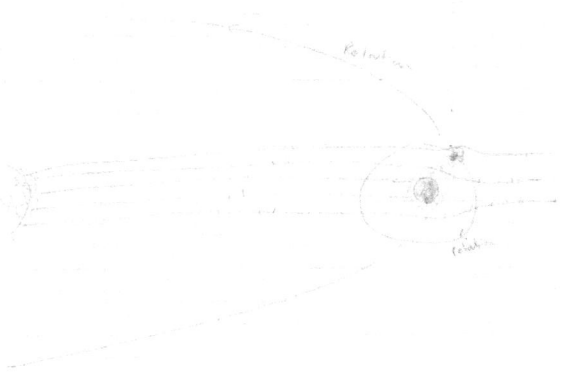

A New Way to See the Universe

Some may argue that my reliance on AI diminishes the legitimacy of my work. However, I see it differently. When scientific calculators were introduced, people resisted. When spell check became standard, skeptics mocked it. However, these tools became essential to progress. AI is no different—it is here to help humanity advance.

**After all, AI is nothing more or less than mathematics and physics.**

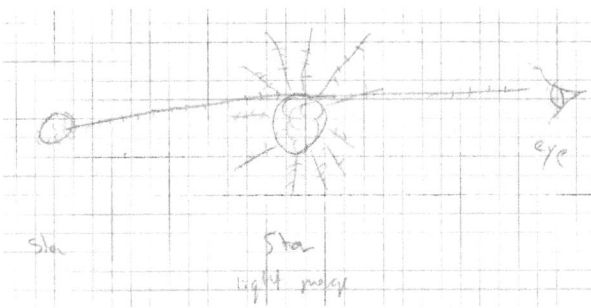

This book is not perfect. It has flaws, and there may be minor, some significant errors. However, it is a start. Every grand Theory begins as an idea, and I hope this work sparks curiosity and discussion. My greatest wish is that the academic community takes this concept, refines it, and presents it for future generations of physicists to explore, challenge, and improve.

I invite you to examine this Theory with an open mind. You need not be a physicist or mathematician to grasp its concepts. You only need curiosity and the willingness to see the universe differently.

Threads and Spinons Theory
Samir Hanna Safar

Threads and Spinons Theory
Samir Hanna Safar

# Part I

# The Origins of the Theory

Threads and Spinons Theory
Samir Hanna Safar

# Chapter 1

## More Historical Context

## The Evolution of Physics and Where It Went Wrong

Before we explore the Threads and Spinons Theory, it is important to understand how modern physics developed and why existing theories fail to fully explain reality. While Newton, Einstein, Bohr, and Heisenberg made remarkable contributions to physics, their frameworks contain flaws, contradictions, and missing pieces that have led to unresolved paradoxes.

This chapter will explore the key historical breakthroughs in physics, highlighting where these theories were incomplete or misguided, setting the stage for the new paradigm of Threads and Spinons.

1. Newton and Classical Physics – The Birth of Force- Based Thinking

Sir Isaac Newton (1643- 1727) laid the foundation for classical mechanics with his laws of motion and law of universal gravitation. His work was revolutionary for its time and correctly described planetary motion, inertia, and forces.

Newton's Key Contributions:

- Gravity as a Universal Force: Newton proposed that all objects attract each other with a force proportional to their masses.

- Laws of Motion: Defined acceleration, momentum, and action- reaction forces.

- Mathematical Precision: Created calculus to describe motion and force dynamics.

Where Newton's Theory Went Wrong:

- No Explanation for the Mechanism of Gravity: Newton described what Gravity does but not why it works.

- Instant Action at a Distance: Gravity in Newton's model acts instantly across space, which contradicts Relativity.

- Space as an Empty Background: Newton assumed space is a passive void, failing to describe any structure in the fabric of the universe.

The Threads and Spinons Theory Corrects This:

- Gravity is not a force acting across space—it is a tension effect in the universal thread network.
- Space is not an empty void—it is filled with structured threads connecting all matter.
- There is no need for instantaneous action— gravitational effects propagate through thread tension dynamics.

2. Einstein and the Warping of Spacetime

Albert Einstein (1879- 1955) revolutionized physics with special Relativity (1905) and general Relativity (1915). He replaced Newton's force- based Gravity with the concept of spacetime curvature.

Einstein's Key Contributions:

- Special Relativity: Explained that time and space are linked, and motion near light speed alters time perception.

- General Relativity: Proposed that mass warps spacetime and this curvature creates what we perceive as Gravity.

- Confirmed by Experiment: Einstein's predictions about gravitational lensing and time dilation were later verified.

Where Einstein's Theory Went Wrong:

- Spacetime is a Mathematical Abstraction: There is no physical explanation for what "curved spacetime" is made of.

- Gravity Still Lacks a Mechanism: Einstein removed Newton's "force" but replaced it with an undefined warping of an imaginary 4D space.

Threads and Spinons Theory
Samir Hanna Safar

- Black Hole Singularities: General Relativity predicts physically impossible infinite- density points.

- Fails to Connect with Quantum Mechanics: Einstein's model does not describe atomic- scale interactions or energy levels.

The Threads and Spinons Theory Corrects This:

- Gravity is not a warping of Spacetime but a result of thread tension pulling objects together.
- There is no need for extra dimensions—only three accurate physical dimensions exist.
- Eliminates black hole singularities—black holes are ultra-dense thread cores, not infinite- density points.

3. Bohr and the Quantum Revolution

Niels Bohr (1885- 1962) played a significant role in early quantum mechanics, introducing the idea of quantized energy levels for electrons inside atoms.

Bohr's Key Contributions:

- Atomic Model: Electrons orbit the nucleus in discrete energy levels.

- Quantum Jumps: Electrons move between levels by absorbing or emitting energy.

- Laid the Foundation for Modern Quantum Mechanics.

Where Bohr's Theory Went Wrong:

- Electrons Do Not Orbit Like Planets: Later discoveries showed electrons behave more like probability waves than defined orbits.

- No Explanation for Why Quantum Levels Exist: Why do electrons only occupy specific states? Bohr's model lacks a deeper explanation.

- Does Not Address the Double- Slit Experiment: The wave-particle duality remains a paradox in this model.

The Threads and Spinons Theory Corrects This:

- Electrons do not "orbit"—they are structured wave patterns of threads and spinons.
- Quantum energy levels arise naturally from thread tension and spinon alignments.
- The double-slit experiment is not about Probability—thread interference patterns cause it.

4. Heisenberg and Quantum Uncertainty

Werner Heisenberg (1901- 1976) introduced the Uncertainty Principle, which states that we can never know both the exact position and velocity of a particle simultaneously.

Heisenberg's Key Contributions:

- Wave- Particle Duality: Showed that particles behave as both waves and particles.

- Quantum Probabilities: Introduced that quantum behavior is governed by Probability, not determinism.

- Foundation for Modern Quantum Mechanics.

Where Heisenberg's Theory Went Wrong:

- Particles Are Not Just "Probability Clouds": The idea that matter is just a set of probabilities lacks a physical mechanism.

- Fails to Explain Quantum Entanglement Mechanically: How can two entangled particles "communicate" instantly across vast distances?

- Does Not Address the Physical Medium for Wave Behavior: What is actually "waving" in quantum waves?

The Threads and Spinons Theory Corrects This:

- Particles are not probability functions but structured formations of threads and spinons.
- Quantum behavior is deterministic at the thread level, not just statistical randomness.
- Entanglement happens because entangled particles remain connected by stretched threads, allowing synchronized spinon motion.

5. The Missing Piece – Why Physics Needs the Threads and Spinons Theory

| Old Physics Models | Threads and Spinons Explanation |
|---|---|
| Gravity is a force (Newton) or a curvature (Einstein). | Gravity is caused by tension in an interconnected thread network. |
| Space is empty, with forces acting through it. | Space is structured—threads fill all of space, guiding energy and matter. |
| Quantum mechanics relies on probability and uncertainty. | Quantum behavior is structured and deterministic within threads. |
| Black holes contain singularities. | Black holes are ultra-dense thread accumulations with defined structure. |
| Consciousness is purely biological. | Consciousness is an organized field of spinon interactions in structured threads. |

The old physics models are incomplete—they describe effects but do not explain the underlying causes.

Threads and Spinons provide the missing link, unifying physics into a single, testable theory.

This theory eliminates paradoxes, <u>replaces abstract math</u> with <u>actual physical structures</u>, and leads to new scientific breakthroughs.

Threads and Spinons Theory
Samir Hanna Safar

Threads and Spinons Theory
Samir Hanna Safar

**Chapter 2**

**Connections to Existing Physics**

**How This Theory Corrects and Expands on General Relativity, Quantum Mechanics, and Electromagnetism**

To fully appreciate the Threads and Spinons Theory, we must address how it connects to and improves upon existing physics. While general Relativity, quantum mechanics, electromagnetism, and neuroscience have provided valuable insights, they also contain gaps and contradictions that have puzzled scientists for decades.

This chapter will bridge the gap between traditional science and the new paradigm, showing how Threads and Spinons integrate and refine key concepts in modern physics.

1. How This Theory Refines General Relativity

- Einstein's General Relativity (GR) describes Gravity as a warping of Spacetime due to Mass.

- It successfully explains planetary motion, time dilation, and gravitational lensing.

- However, it fails at singularities, cannot explain dark matter/energy, and does not unify with quantum mechanics.

Threads and Spinons Correct These Issues:

- Gravity is not a "curvature" of Spacetime—it results from tension in a structured thread network.

- No need for extra dimensions—Gravity is a three-dimensional tension phenomenon.

- Eliminates singularities—black holes are ultra-dense thread cores, not infinitely small points.

Key Difference:

- General Relativity describes Gravity using equations but does not explain its physical nature.
- Threads and Spinons provide a physical mechanism—Gravity is an absolute force exerted by stretched threads, not an abstract mathematical effect.

2. How This Theory Fixes Quantum Mechanics' Inconsistencies

- Quantum mechanics explains the atomic and subatomic world but relies on Probability and uncertainty.

- The wave-particle duality and Entanglement remain poorly understood.

- Quantum mechanics contradicts Relativity—there is no unified framework.

Threads and Spinons Provide a Physical Explanation:

- Particles are not just probability waves but structured formations within a real thread network.

- Wave-particle duality occurs because spinons move along threads, creating interference patterns.

- Quantum Entanglement is not "spooky action"—entangled particles remain physically connected by stretched threads.

Key Difference:

- Quantum mechanics treats probabilities as fundamental.
- Threads and Spinons treat physical structure as fundamental—quantum behavior is structured, not random.

3. How This Theory Reinterprets Electromagnetism

- Maxwell's equations describe how electric and magnetic fields behave but do not explain what they physically are.

- Electromagnetic waves are assumed to travel through "space" without a medium.

- Charge interactions (Coulomb's law) describe forces between charges but do not explain why they occur.

Threads and Spinons Provide a Physical Explanation:

- Electric and magnetic fields arise from spinon motion along conductive threads.

- light is not just a wave but a structured disturbance propagating through threads.

- Charge interactions result from thread alignments between particles, not abstract force exchanges.

Key Difference:

- Electromagnetism describes field behavior but does not define its underlying mechanism.
- Threads and Spinons define electromagnetism as structured spinon interactions within the thread network.

4. How This Theory Expands Neuroscience and Consciousness Studies

- Modern neuroscience assumes the brain generates Consciousness purely through neurons and synapses.

- However, memories persist even when neurons die, and consciousness phenomena like déjà vu and intuition remain unexplained.

- The brain is treated as an isolated system without accounting for non-local connections or quantum effects.

Threads and Spinons Theory
Samir Hanna Safar

Threads and Spinons Provide a New Model of Consciousness:

- Thoughts and memory are structured as spinon imprints in neural thread networks.

- Consciousness extends beyond the brain—spinons allow non- local interactions, explaining intuition and subconscious thought.

- Memory retrieval is a resonance effect, not a static neural storage mechanism.

Key Difference:

- Traditional neuroscience focuses on chemical and electrical signals in neurons.
- Threads and Spinons explain how Consciousness is a structured energy field that interacts with thread networks.

5. How This Theory Resolves the Dark Matter and Dark Energy Problem

- Dark matter and dark energy are placeholders for unknown forces in cosmology.

- Dark matter is inferred because galaxies rotate faster than expected, but no particle has been detected.

- Dark energy is assumed to drive universal expansion, yet its nature is unknown.

Threads and Spinons Provide a New Explanation:

- Dark matter is not a mysterious particle—it is an effect of unseen thread density affecting Gravity.

- Dark energy is not a separate force but a natural effect of universal thread expansion.

Key Difference:

- Traditional physics adds new invisible entities to explain missing Mass and energy.
- Threads and Spinons show how these effects arise naturally from thread structures.

6. How This Theory Unifies All Forces into One Model

Current Understanding of Fundamental Forces:

| Force | Traditional Explanation | Threads and Spinons Explanation |
|---|---|---|
| Gravity | Mass warps spacetime (General Relativity) | Tension in threads pulls masses together |
| Electromagnetism | Mediated by photons (Maxwell's Equations) | Spinon movement along conductive threads |
| Strong Nuclear Force | Gluons hold quarks together | Nuclear bonds are formed by tightly compressed thread loops |
| Weak Nuclear Force | W and Z bosons cause decay | Weak interactions are caused by thread reconfiguration |

- All forces emerge from the fundamental structure of threads and spinons, not separate particles.

- This eliminates force- carrying particles (gravitons, gluons, W/Z bosons) and replaces them with a single unified mechanism.

7. Why This Theory Is More Testable Than String Theory?

Threads and Spinons Theory
Samir Hanna Safar

- String theory attempts to unify physics but is based on unobservable extra dimensions.

- It makes no direct testable predictions and is essentially a mathematical construct.

- It does not explain Gravity mechanistically—only as a side effect of higher- dimensional vibrations.

Threads and Spinons Provide a Testable Alternative:

- It describes a physical mechanism for Gravity, electromagnetism, and quantum behavior.

- It proposes experiments that measure thread structures (light-speed variations, thread-induced gravity effects, spinon communication).

- It does not require unobservable dimensions—only actual, physical threads in 3D space.

Key Difference:

- String theory relies on abstract mathematical models without experimental support.
- Threads and Spinons offer concrete, testable predictions based on fundamental physical structures.

Conclusion:

A New Scientific Framework that Unifies and Expands Physics

The Threads and Spinons Theory accepts previous physics—it builds upon and corrects its limitations by providing a physical explanation for phenomena previously described only mathematically.

This theory:

Explains Gravity as thread tension, not Spacetime warping. Replaces quantum randomness with structured spinon behavior. Redefines electromagnetism as spinon flow along conductive threads. Unites fundamental forces into a single framework without force-carrying particles. Explains Consciousness as structured energy fields interacting with the universe.

## Chapter 3

### Addressing Objections

### Why Mainstream Physics Falls Short and How This Theory Fills the Gaps

Any new theory faces skepticism—especially when it challenges established physics. The Threads and Spinons Theory offers a fundamental shift from mainstream concepts, and it is natural for scientists to question its validity.

This chapter will address key objections from physicists and explain why alternative theories, such as General Relativity, Quantum Mechanics, and String Theory, fail to describe reality fully.

1. "Why Should We Abandon General Relativity?"

Objection: "Einstein's General Relativity has been tested and confirmed through experiments like gravitational lensing, time dilation, and GPS accuracy. Why should we replace it?"

Response:

- General Relativity accurately describes Gravity's effects but does not explain its cause.

- It requires spacetime curvature but provides no physical mechanism for how mass "tells" space to bend.

- Singularities at black hole centers are mathematical artifacts that break physics.

Why General Relativity Fails:

- Spacetime curvature is a mathematical concept, not a physical entity.

- Cannot explain quantum- scale Gravity.

- Requires dark matter and dark energy to fill in missing gaps.

How Threads and Spinons Fix This:

- Gravity is caused by fundamental physical thread tension, not abstract curvature.

- Eliminates singularities by defining black holes as dense thread <u>cores</u>.

- No need for dark energy—cosmic expansion is caused by thread growth.

2. "Quantum Mechanics is the Most Accurate Model We Have—Why Replace It?"

Objection: "Quantum mechanics has never been wrong in an experiment. It is the foundation of all modern technology. Why change it?"

Response:

- Quantum mechanics works mathematically but does not explain the mechanisms behind its effects.

- It relies on Probability instead of physical causality.

- Wave- particle duality, Entanglement, and uncertainty are paradoxes that remain unresolved.

Why Quantum Mechanics Fails:

- Probability is a shortcut for missing knowledge—not an explanation.

- Quantum mechanics cannot explain why wave functions collapse.

- Entanglement appears "spooky" because it lacks a medium for connection.

How Threads and Spinons Fix This:

- Wave- particle duality is resolved—particles are structured spinon interactions along threads.

- Entanglement happens because particles remain physically connected through stretched threads.

Threads and Spinons Theory
Samir Hanna Safar

- Quantum randomness is not fundamental but a side effect of unobserved thread dynamics.

3. "If This Theory is Correct, Why Has No One Detected Threads?"

Objection: "If the universe is made of threads, why haven't we seen them in experiments?"

Response:

- Threads are extremely thin and exist at subatomic scales, making them difficult to detect with current instruments.

- Gravitational and electromagnetic forces suggest an underlying medium, but physics has been searching in the wrong direction (e.g., gravitons, dark matter).

- The effects of threads are observable (Gravity, light behavior, Entanglement), even if they are hard to measure directly.

How to Test for Threads:

- Proposed Experiments:

  - Light- speed anisotropy tests—tiny variations in speed should appear if light moves through threads.
  - Gravity propagation experiments—gravitational waves should show subtle deviations if threads transmit force.
  - Electromagnetic resonance experiments—spinon interactions in threads should produce unique quantum signatures.

Just as we inferred atoms long before seeing them, we must detect threads indirectly before direct confirmation.

4. "Doesn't This Sound Like String Theory?"

Objection: "String theory already proposes fundamental strings. How is this different?"

Response:

- String theory requires extra dimensions (10- 26) unobservable and purely theoretical.

- It does not provide testable predictions—no string has ever been detected.

- It does not explain Gravity mechanistically, only as a side effect of higher- dimensional vibrations.

Why String Theory Fails:

- It is mathematically elegant but has no experimental proof.

- It does not unify physics in a testable way.

- It relies on hidden dimensions that cannot be observed or measured.

How Threads and Spinons Fix This:

- Only requires three dimensions—no extra dimensions needed.

Threads and Spinons Theory
Samir Hanna Safar

- Makes testable predictions about Gravity, light, and quantum mechanics.

- Defines actual physical structures (threads and spinons) instead of abstract mathematical objects.

5. "What About the Higgs Boson and the Standard Model?"

Objection: "The Higgs boson was discovered, confirming the Standard Model. Why change our understanding of Mass?"

Response:

- The Higgs boson explains how particles acquire Mass but does not explain why Mass creates Gravity.

- The Higgs field is another abstract mathematical concept—it does not provide a physical structure.

- Why the Standard Model Fails:

- It requires 17+ fundamental particles but cannot explain why they exist.

- Does not include Gravity—General Relativity is still separate.

- Relies on invisible quantum fields instead of physical mechanisms.

How Threads and Spinons Fix This:

- Mass arises from thread interactions, not an abstract field.

Threads and Spinons Theory
Samir Hanna Safar

- Gravity and electromagnetism emerge naturally from thread structures.

- Unifies forces instead of treating them as separate phenomena.

6. "Why Hasn't Any Major Physicist Proposed This Before?"

Objection: "If this theory is correct, why hasn't the scientific community accepted it?"

Response:

- Physics is highly resistant to paradigm shifts—just as heliocentrism, quantum mechanics, and Relativity faced opposition.

- Modern physics relies on mathematical models that work well in limited, incomplete domains.

- Breakthroughs often come from outside the establishment— Einstein was a patent clerk when he rewrote physics.

Why the Scientific Community Resists Change:

- Established theories are deeply entrenched in academia.
- Funding and research depend on supporting existing models.
- Significant discoveries require experimental confirmation, which takes time.

This is not the first time a new theory has faced resistance— but experimental evidence will ultimately decide.

Threads and Spinons Theory
Samir Hanna Safar

## 7. Why This Theory is the Best Path Forward

| Objection | Response from Threads and Spinons |
| --- | --- |
| "General Relativity works!" | Yes, but it does not explain why gravity exists—it only describes effects. |
| "Quantum mechanics is proven!" | Yes, but it lacks a physical mechanism—probabilities do not explain structure. |
| "Why haven't we seen threads?" | We detect their effects (gravity, entanglement) and can test them experimentally. |
| "Isn't this just String Theory?" | No—this requires only 3 dimensions and is experimentally testable. |
| "The Standard Model is complete!" | No—mass is still unexplained, and gravity is not included. |
| "Why hasn't anyone proposed this before?" | Science resists paradigm shifts—new discoveries take time to gain acceptance. |

This theory is built on testable predictions. It eliminates paradoxes rather than adding complexity. It provides an accurate physical framework for Gravity, quantum mechanics, and electromagnetism.

# Part II

# The Core Scientific Framework

Threads and Spinons Theory
Samir Hanna Safar

**Chapter 4**

## Light – A Thread- Based Phenomenon

### Redefining Light: A Continuous Thread Model

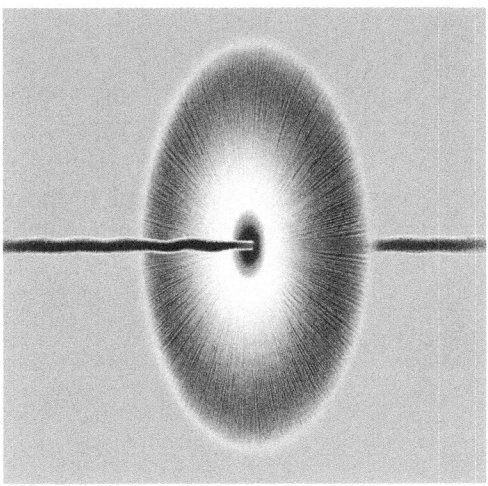

In modern physics, Light is often described as both a particle (photon) and a wave (electromagnetic radiation). Though widely accepted, this duality has led to paradoxes and unresolved questions. The Threads and Spinons Theory presents an entirely different approach:

Light is not a particle. Light is a continuous thread—a physical structure stretching through space, carrying energy through the motion of spinners.

Rather than existing as discrete photons traveling through emptiness, Light results from disturbances propagating along an interconnected network of threads. These threads are fundamental, forming the structure of reality itself, and light moves along them as a disturbance, much like vibrations traveling through a taut string.

The Structure of Light in the Threads and Spinons Theory

1. Threads:

> Light is composed of a single, continuous thread extending through space.

> This thread does not break or move like a particle but instead vibrates and transfers energy.

2. Spinons (Rotating Energy Carriers):

> Spinons are rotational energy elements embedded within threads.

> They rotate at specific frequencies, determining the characteristics of the light wave.

> Higher frequency spinon rotations correspond to higher- energy Light (e.g., ultraviolet, gamma rays).

3. Wave Behavior as a Thread Vibration:

> A "wave" in classical physics is a disturbance traveling through the thread.

Threads and Spinons Theory
Samir Hanna Safar

The amplitude of this vibration determines the intensity of Light.

The frequency of the Spinon's rotation within the thread determines the color (for visible Light) or type (infrared, X- ray, etc.).

**Key Differences from Modern Physics**

| Modern Physics (Photon Model) | Threads and Spinons Theory |
| --- | --- |
| Light is made of **particles (photons)**. | Light is a **continuous thread disturbance**. |
| Light exhibits **wave-particle duality**. | Light behaves purely as a **thread vibration**. |
| Photons travel as discrete packets. | Light propagates as **spinon-induced thread waves**. |
| Spacetime dictates light's movement. | **Thread tension determines light's speed.** |
| Light can be absorbed/emitted in quantized amounts. | **Spinon energy transfer defines absorption and emission.** |

The Double-Slit Experiment Explained

The double- slit experiment has long been a mystery in quantum mechanics, showing that Light behaves as both a particle and a wave. According to the Threads and Spinons Theory:

- Before observation: Light exists as a thread-based wave disturbance. As it passes through the slits, it follows multiple possible paths simultaneously, just like a tensioned string oscillating in multiple modes.
- During measurement (observation): The observation disrupts the natural thread alignment, forcing the light wave to collapse into a defined path, creating the appearance of a particle-like photon.

Threads and Spinons Theory
Samir Hanna Safar

Thus, the so-called "wave-function collapse" is simply the forced reconfiguration of the threads upon interaction with a measuring device. This provides a logical, physical explanation for what is usually considered a quantum mystery.

Entanglement Explained: A Shared Thread Connection

Quantum entanglement is another mystery of modern physics, where two photons remain connected regardless of distance. In the Threads and Spinons framework:

- Two entangled "photons" are sections of the same continuous thread.
- Any disturbance in one section of the thread immediately affects the other because the thread itself is a single entity.
- This eliminates the need for faster- than-light information transfer—the information is already present within the thread.

This resolves the paradox of quantum non-locality while maintaining the fundamental realism of physics.

The Speed of Light: A Property of Thread Tension

In relativity, the speed of light (c) is treated as a universal constant independent of its medium. However, in the Threads and Spinons Theory, the speed of Light is a direct result of thread tension:

in the Threads and Spinons Theory, the speed of light is a direct result c

$$c = \sqrt{\frac{T}{\mu}}$$

Where:

- $T$ = Thread Tension (a fundamental property of space)
- $\mu$ = Linear mass density of the thread

The speed of Light is not an arbitrary constant but a function of the physical properties of the thread structure itself. This model explains why Light slows down in different media—the effective thread tension changes due to interactions with atomic structures.

Absorption and Emission of Light

Light is absorbed and emitted in quantum mechanics as discrete packets (quanta). In the Threads and Spinons Theory, this process is explained through spinon energy transfer:

1. Absorption: When an atom absorbs energy, it excites the spinons within its threads, increasing their rotational speed.
2. Emission: When the atom loses energy, the spinons release their rotational energy back into the thread, creating a new propagating wave of Light.

This mechanism eliminates the need for "photon particles" while preserving the quantized nature of energy exchange.

Redefining Blackbody Radiation

The phenomenon of blackbody radiation—the emission of Light from heated objects—is also a direct result of thread-based energy transfer:

- Higher temperatures increase spinon rotational speeds, shifting emissions toward higher- frequency Light (e.g., infrared to visible to ultraviolet).
- The classical "ultraviolet catastrophe" is avoided because spinon energy transfer follows a natural tension- based distribution, not an infinite summation of modes.

Conclusion:

A Unified View of Light

The Threads and Spinons Theory redefines Light as:

1. A continuous thread structure eliminates the need for photons.
2. Spinon- based energy transfer, replacing quantum uncertainty with mechanical predictability.
3. A vibration- dependent speed, removing arbitrary constants and connecting Light's properties to fundamental physical laws.
4. A natural explanation for quantum effects, resolving paradoxes like the double- slit experiment and entanglement.

This Theory brings Light back into tangible, physical interactions, eliminating unnecessary abstractions while preserving all observed behaviors.

With this new foundation, we can explore more profound aspects of physics—how matter interacts, how galaxies form, and how energy propagates throughout the universe.

**This is just the beginning.**

Threads and Spinons Theory
Samir Hanna Safar

**Chapter 5**

## The Light Spectrum

### How Spinon Stacks Define Electromagnetic Waves

Light has been one of the most mysterious and debated phenomena in physics. Traditional physics explains Light as both a wave and a particle, leading to wave- particle duality—a concept that remains unsatisfactorily explained.

The Threads and Spinons Theory offers a new framework in which Light is not an abstract wave or particle but a structured phenomenon of spinon motion along threads.

In this chapter, we will define the full electromagnetic spectrum and explain how variations in spinon stacks and spinon rotational speed cause different wavelengths.

What Is Light? Redefining Electromagnetic Waves

Traditional Explanation:

- Electromagnetic waves consist of oscillating electric and magnetic fields that propagate through space.
- Quantum physics says Light is a particle (photon), creating wave- particle duality.
- No physical mechanism is given for how "waves" or "fields" exist in a vacuum.

Problems with the Traditional View:

- Waves require a medium, but Light supposedly propagates through "space."

- Photons are treated as both particles and waves, an unresolved paradox.

- Electromagnetic waves are mathematical constructs without a defined physical structure.

Threads and Spinons Explanation:

- Light is a structured energy pulse traveling along interconnected cosmic threads.

- Spinons (rotating energy units) travel along these threads, creating wave- like effects.

- The speed, frequency, and energy of Light depend on the number of stacked spinons and their rotation rates.

Threads and Spinons Theory
Samir Hanna Safar

$$E_{\text{light}} = S_{\text{spinon}} \times N_{\text{stack}} \times SEU$$

where:

- $E_{\text{light}}$ = Energy of the light wave
- $S_{\text{spinon}}$ = Spinon rotational speed
- $N_{\text{stack}}$ = Number of spinons stacked together
- $SEU$ = Spinon Energy Unit

- No need for an electromagnetic "field"—Light is the direct movement of structured spinon waves.

- Explains wave and particle behavior—spinon stacks behave like discrete units yet propagate like waves.

How Spinon Stacks Define the Light Spectrum

The electromagnetic spectrum is defined by wavelength and frequency, ranging from low- energy radio waves to high-energy gamma rays. However, why do these differences exist?

In the Threads and Spinons Theory:

- Each part of the spectrum is determined by the number of spinons in a stack and their rotational velocity.

- Higher- energy waves have more stacked spinons rotating at higher speeds.

- Lower- energy waves have fewer stacked spinons or slower spinon motion.

Defining the Spectrum Based on Spinon Stacks:

| Wave Type | (m) | $N_{stack})$ | $S_{spinon})$ | Energy (SEU) |
|---|---|---|---|---|
| Radio Waves | $> 10^{1}$ | 1-2 spinons | Slow rotation | Low energy |
| Microwaves | $10^{-3}$ | 2-5 spinons | Moderate rotation | Slightly higher |
| Infrared (IR) | $10^{-6}$ | 5-10 spinons | Faster rotation | Medium energy |
| Visible Light | $10^{-7}$ | 10-15 spinons | High-speed rotation | High energy |
| Ultraviolet (UV) | $10^{-8}$ | 15-20 spinons | Very high rotation | Very high energy |
| X-Rays | $10^{-10}$ | 20-30 spinons | Extreme rotation | Intense energy |
| Gamma Rays | $10^{-12}$ | 30+ spinons | Max rotation speed | Maximum energy |

- Radio waves contain only a few spinons rotating at low speeds.

- Visible Light is produced when 10- 15 spinons rotate in a synchronized stack.

- Gamma rays have the most stacked spinons, spinning at maximum energy capacity.

Key Takeaways:

- Higher- energy Light is a denser stack of spinning spinons traveling along threads.

- No need for "photon particles"—Light's energy level is fully explained by spinon stacking.

- Electromagnetic waves are not oscillating fields but structured motion patterns within thread networks.

How Spinon Rotation Determines Light Frequency

What Determines Light's Frequency?

Threads and Spinons Theory
Samir Hanna Safar

Traditional View: Frequency is just the number of oscillations per second.

Spinon Theory View: Frequency results from spinon rotation speed and their synchronized movement within a stack.

$$f_{\text{light}} = \frac{S_{\text{spinon}}}{N_{\text{stack}}}$$

where:

- $f_{\text{light}}$ = Frequency of the wave
- $S_{\text{spinon}}$ = Spinon rotational velocity
- $N_{\text{stack}}$ = Number of spinons in the stack

- Faster spinon motion creates higher frequencies.

- More stacked spinons lower the frequency but increase energy.

This explains why visible Light has a narrow frequency range—human vision is tuned to a specific spinon rotation window.

Why Light Moves at a Constant Speed

- Traditional physics assumes that the speed of Light (ccc) is a universal constant but does not explain why.

- The Threads and Spinons Theory explains why Light moves at the speed it does.

Spinon Motion Determines Light Speed

Threads and Spinons Theory
Samir Hanna Safar

- Spinons travel along pre- existing threads, like beads moving on a string.

- The speed of Light is a function of the tension and density of the threads.

$$c = \frac{T_{\text{thread}}}{\rho_{\text{thread}}}$$

where:

- $c$ = Speed of light
- $T_{\text{thread}}$ = Thread tension
- $\rho_{\text{thread}}$ = Thread density in a given region

Why is this important?

- In regions of space with higher thread density, Light may travel slightly slower.
- This predicts subtle variations in light speed that could be tested experimentally.
- It eliminates the need for special relativity's postulate that light speed is an unexplained universal constant.

- Light speed is determined by the properties of the threads—not an arbitrary law of nature.

Practical Applications of This Model

Revolutionizing Communication:

- If we can manipulate spinon stacks, we can control and enhance light transmission, leading to faster-than- light spinon- based communication.

New Energy Technologies:

- Controlling spinon rotation can generate energy directly from structured light waves, opening new possibilities for quantum energy extraction.

Reinterpreting Astrophysical Phenomena:

- Light bending (gravitational lensing) may not be due to "spacetime warping" but rather variations in thread tension affecting spinon movement.
- Cosmic background radiation could result from ancient thread configurations, not leftover Big Bang energy.

Understanding light at the spinon level could lead to groundbreaking technologies in optics, communication, and energy.

Conclusion:

A New Definition of Light

| Traditional Model | Threads and Spinons Explanation |
| --- | --- |
| Light is an electromagnetic wave | Light is structured spinon motion along threads |
| Light is also a particle (photon) | Light appears quantized due to spinon stacking |
| Frequency is an abstract number | Frequency depends on spinon speed and stack count |
| Speed of light is a universal constant | Light speed is determined by thread tension and density |

Threads and Spinons Theory
Samir Hanna Safar

Light is not just an abstract electromagnetic field but a structured spinon motion pattern traveling through the universe's thread network.

This explains wave- particle duality, the speed of Light, and the entire electromagnetic spectrum in a single model.

**Chapter 6**

**The Six Fundamental Laws of the Threads and Spinons Theory**

In modern physics, laws are built upon complex mathematical models that often require unobservable assumptions. The Threads and Spinons Theory simplifies these concepts by introducing six fundamental laws that govern the universe. These laws describe the behavior of threads (the foundational structure of reality) and spinons (rotational energy carriers), providing a unified framework for understanding matter, Light, gravity, and energy transfer.

Law 1: The Law of Continuous Threads

All particles and forces in the universe are composed of fundamental **threads**, which extend continuously and interweave to form all structures.

This law replaces the concept of point-like particles with a network of unbroken threads that shape reality. Instead of discrete particles moving through space, the universe consists of an intricate web of interacting threads. Atoms, photons, and even gravitational fields emerge from the dynamic interactions of these threads, defining the fundamental nature of existence.

Implications:

- Matter is not made of independent particles but is a manifestation of threads configurations.
- Energy is transferred through the vibrations and rotations of the thread.
- There is no "empty space"—threads are interconnected.

Law 2: The Law of Spinon Motion

Spinons are the fundamental units of motion and energy transfer. Their rotational speed determines the frequency, and their stacking determines energy.

Spinons are rotating elements within the thread. Unlike traditional physics, where energy is stored in abstract fields or virtual particles, energy exists as rotational motion in spinons.

Implications:

- The faster a spinon rotates, the higher the wave frequency it generates (e.g., visible Light vs. X- rays).
- Energy is not quantized in the sense of discrete photons but rather as stacked spinons—more stacks create more energy.
- Heat, magnetism, and electromagnetism all emerge from spinon interactions.

Law 3: The Law of Thread Tension and Wave Propagation

The speed of any wave (Light, sound, or gravity) is determined by the tension and mass density of the thread.

Instead of treating the speed of Light as an arbitrary constant, this law states that speed is a physical property of the thread itself:

$$c = \sqrt{\frac{T}{\mu}}$$

Where:

- $T$ = Thread Tension (a fundamental property of space)
- $\mu$ = Linear mass density of the thread

Implications:

- The speed of Light is not an absolute universal constant but a result of thread properties.
- Changes in thread tension due to external influences (e.g., gravitational fields) can affect the speed of Light.

- Gravity, sound, and electromagnetic waves all obey the same fundamental rule of wave propagation.

Law 4: The Law of Thread Bonding and Attraction

Objects attract each other due to thread tension, not spacetime curvature or particle exchange.

Traditional physics explains gravity using curved spacetime or force- carrying particles (gravitons). The Threads and Spinons Theory eliminates both concepts and states that gravity results from thread contraction and tension.

Implications:

- Gravity is not a force but a tension effect—denser masses pull on surrounding threads, creating attraction.
- This explains why all objects fall at the same Rate in a vacuum (Galileo's falling bodies experiment).
- Black holes (now called Cores) are dense thread bundles, not singularities.

Law 5: The Law of Energy Transfer Through Threads

Energy moves through threads, and its intensity depends on the rotational strength of spinons.

This law unifies different types of energy transfer, including heat, electromagnetism, and nuclear forces, under one principle: all energy travels via thread motion.

Implications:

- Heat is the transfer of spinon rotation from one region to another.
- Electromagnetism is a structured spinon interaction that aligns threads into magnetic fields.
- Nuclear energy is stored within the thread configurations of atoms, not within "strong" and "weak" forces.

Law 6: The Law of Thread Expansion and Universal Growth

The universe did not originate from a singularity (Big Bang). Instead, it continuously expands as threads stretch and spinons multiply.

Instead of an explosive beginning, the universe grows over time as vacuum pressure causes threads to expand and generate more spinons.

$$U(t) = U_0 + \int_0^t R\,dt$$

Where:

- $U(t)$ = Universe size at time $t$
- $U_0$ = Initial size of the universe
- $R$ = Rate of thread expansion

Implications:

- The universe has no beginning or end—it is in a constant state of growth.
- Dark energy is an illusion—what we observe as cosmic expansion is a thread stretching over time.

Threads and Spinons Theory
Samir Hanna Safar

- Galaxy formation is driven by Core growth, not singularity remnants.

Conclusion:

A New Set of Laws for Physics

These six laws redefine physics, replacing abstract concepts with physical, testable mechanisms. Instead of relying on invisible dimensions, spacetime curvature, or force carriers, the Threads and Spinons Theory offers a unified, realistic framework that explains the universe directly and mechanically.

This is not just a modification of existing physics but a fundamental rethinking of how reality operates.

Threads and Spinons Theory
Samir Hanna Safar

**Chapter 7**

**The 20 Principles of the Threads and Spinons Theory**

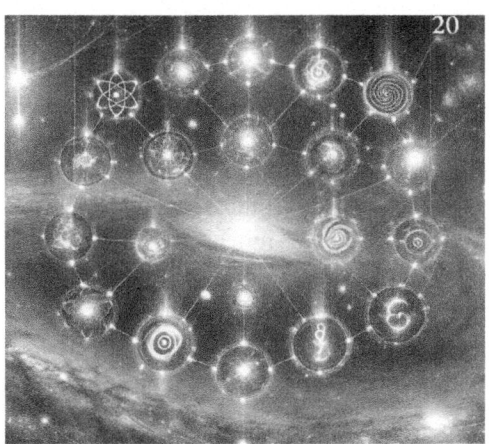

The Threads and Spinons Theory introduces 20 fundamental principles that govern the physical universe. These principles challenge mainstream physics by removing the reliance on unobservable dimensions, spacetime curvature, and force- carrying particles. Instead, they offer a mechanical, interconnected framework where everything arises from threads and spinons.

These principles support and expand upon the six fundamental laws outlined in the previous chapter.

1. The Universe is Made of a Single Continuous Thread

Threads and Spinons Theory
Samir Hanna Safar

- Everything in the universe—matter, energy, and forces—comes from an infinitely long, continuous thread woven into structures.
- This thread does not break; it loops, folds, and forms all known particles and waves.

2. Spinons are the Fundamental Carriers of Energy

- Energy is not stored in abstract fields or force carriers but in spinons—rotational disturbances within the thread.
- The faster a spinon rotates, the higher its energy (e.g., X- rays vs. visible Light).

3. Light is Not a Particle; It is a Thread Vibration

- No photons exist—Light is a wave disturbance traveling along a thread.
- The wave's frequency is dictated by spinon motion within the thread.

4. Wave- particle duality is an Illusion

- The double- slit experiment does not show photons acting as particles and waves.
- Instead, Light's thread- based interference patterns create wave- like behavior.

5. Gravity is the Effect of Thread Contraction and Tension

- Objects do not warp spacetime.
- Gravity arises when mass pulls in surrounding threads, creating tension that pulls other objects inward.

## 6. The Speed of Light is a Function of Thread Tension

- Light's speed is not an arbitrary constant but depends on the thread's tension and density:

$$c = \sqrt{\frac{T}{\mu}}$$

- Due to thread properties, Light can slow down in different environments (near stars, black holes, or dense materials).

## 7. Matter is a Stable Configuration of Threads

- Atoms and subatomic particles are not discrete objects but stable thread loops and knots.
- Protons, neutrons, and electrons are structured thread formations held together by spinon motion.

## 8. Electromagnetism is a Spinon Alignment Effect

- Magnetic fields arise from spinons aligning in threads, not from abstract field lines.
- Electricity is the controlled movement of spinons along thread networks.

## 9. Heat is a Transfer of Spinon Rotational Energy

- Heat is not random particle motion but the rotational energy transfer between spinons.

- Temperature measures the intensity of spinon rotation within a thread.

10. The Universe is Expanding Due to Thread Stretching, Not the Big Bang

- The universe did not start from a singularity.
- Instead, threads continuously stretch, producing more spinons, leading to cosmic expansion.

$$U(t) = U_0 + \int_0^t R\,dt$$

- This explains why galaxies appear to be moving apart—thread expansion causes universal growth.

11. Quantum Uncertainty is an Observer Effect, Not a Fundamental Property

- Quantum mechanics suggests that particles have uncertain positions.
- In this Theory, "uncertainty" happens because measuring a thread disturbs it, affecting spinon motion.

12. There are No Force- Carrying Particles

- No gravitons, no gluons, no W/Z bosons.
- Forces arise from thread interactions, not from virtual particles.

## 13. Atomic Stability is Due to Thread Tension, Not the Strong Force

- The so- called "strong force" does not require force-carrying particles.
- Instead, protons and neutrons are stabilized by thread tension and spinon rotation.

## 14. Quantum Entanglement is a Shared Thread Effect

- Entangled particles are sections of the same continuous thread.
- When one changes, the other responds because they are physically linked, not because of faster- than-light communication.

## 15. There is No Dark Matter—Gravity Behaves Differently at Cosmic Scales

- The "missing mass" in galaxies is not caused by invisible dark matter.
- Instead, gravitational pull weakens differently due to thread structure at large scales.

## 16. Black Holes Are Not Singularities—They Are Dense Cores

- Black holes (Cores) are ultra- compressed thread structures, not singularities with infinite density.
- They absorb threads, and when critical density is reached, they explode to form new galaxies.

## 17. The Neutrino is a Spinon-Based Wave Disturbance, Not a Particle

- Neutrinos are not tiny particles traveling through space.
- They are low-energy wave ripples moving along threads, undetectable except in specialized experiments.

18. Matter-Antimatter Annihilation is a Thread Unraveling Process

- When matter meets antimatter, their opposing thread configurations unwind, releasing stored spinon energy.
- No particles "disappear"—threads return to their lower- energy state.

19. Time is Not a Fourth Dimension—Only Three Dimensions Exist

- Time is not a fundamental property—it is our perception of thread interactions and energy transfer.
- Events do not "travel" through time; they change based on thread tension and spinon rotation.

20. All Forces Are Variations of the Same Thread Interactions

- Gravity, electromagnetism, nuclear forces, and weak interactions are not separate.
- They are all emergent effects of how threads interact and store spinon energy.

Conclusion:

A Framework Rooted in Reality

Threads and Spinons Theory
Samir Hanna Safar

These 20 principles define a new framework for physics, eliminating unnecessary complexities and providing an accurate, testable foundation for how the universe works.

\

Threads and Spinons Theory
Samir Hanna Safar

**Chapter 8**

## The Spinon Energy Unit (SEU)

## A New Measure of Energy

One of the most significant breakthroughs in the Threads and Spinons Theory is introducing a new unit of measure for energy: the Spinon Energy Unit (SEU). Traditional physics uses units such as joules (J) and electron volts (eV) to measure energy. However, these units are based on classical mechanics and quantum assumptions that do not align with the thread- based structure of the universe.

The SEU was developed to provide a natural, intuitive, and fundamental unit of energy directly tied to the rotational motion of spinons within the universal thread structure.

1. Why Was the SEU Needed?

Problems with Traditional Energy Units:

Joules (J) are defined using macroscopic mechanical principles, which are unrelated to the fundamental nature of energy in the Threads and Spinons Theory.

Electron volts (eV) rely on an electron's charge, which assumes a particle- based model of matter that this Theory rejects.

Quantum mechanics treats energy as quantized (photon packets), whereas the Threads and Spinons Theory describes energy as continuous rotational motion along a thread.

To resolve these issues, we needed a unit that:

- Directly measures the rotational energy of spinons

- Relates to thread tension and wave motion

- Can be used across gravitational, electromagnetic, and quantum scales

2. The Definition of SEU

The Spinon Energy Unit (SEU) is defined as:

Threads and Spinons Theory
Samir Hanna Safar

$$1 \text{ SEU} = k_s \cdot f_s^2$$

Where:

- $k_s$ = **Thread Tension Constant** (determined experimentally)
- $f_s$ = **Spinon rotational frequency** (measured in cycles per second)

This equation shows that energy is directly proportional to the square of spinon rotational frequency. Higher spinon rotation means more stored energy.

How SEU Relates to Other Units:

Although the SEU is fundamentally different from traditional units, it can be approximately converted into classical terms for comparison:

$$1 \text{ SEU} \approx 0.02 \text{ Joules}$$

$$1 \text{ SEU} \approx 1.25 \times 10^{18} \text{ eV}$$

This conversion allows us to test SEU- based predictions against known experimental data.

3. How SEU Was Developed in This Theory

The need for SEU became apparent when we attempted to describe gravitational and electromagnetic interactions without conventional quantum energy quantization.

We started by:

Threads and Spinons Theory
Samir Hanna Safar

1. Defining energy as a function of thread properties (instead of using photons or virtual particles).
2. Examining energy transfer in Light, heat, and gravity through the behavior of spinons.
3. Finding a mathematical relationship between spinon rotation and energy output, leading to the equation:

$$E = k_s \cdot f_s^2$$

4. Testing predictions with real- world observations (e.g., comparing LIGO gravitational wave data to SEU- based calculations).

Through this approach, we arrived at a consistent unit of energy that works across all scales of physics—from quantum interactions to galaxy formation.

4. SEU in Practical Applications

1. Light and Electromagnetic Radiation

- Instead of photons, light energy is measured in SEUs based on spinon frequency.
- Infrared, visible, and X- rays are different SEU spinon activity levels within the thread.

2. Gravity and Mass- Energy Equivalence

- The famous equation $E = mc^2$ is redefined using SEU:

$$E = mT \cdot SEU$$

- Mass is a function of thread configuration, and SEU provides the energy stored within massive objects.

3. Cosmic Expansion and Thread Growth

- The universe expansion rate is measured in SEUs, replacing dark energy assumptions.
- SEU explains galactic energy distributions without requiring mysterious "dark matter."

4. Nuclear Energy and Fusion

- The energy released in fusion reactions is not due to particles' binding energy but SEU- based spinon realignment within atomic threads.
- This new perspective could lead to more efficient fusion energy designs.

5. The SEU and Future Experiments

To validate SEU as a fundamental unit of energy, we propose experiments that:

- Measure thread tension in different energy interactions

- Compare SEU predictions to LIGO gravitational wave observations

- Analyze light emission without assuming photon quantization

Threads and Spinons Theory
Samir Hanna Safar

- Test alternative fusion models based on thread- spinon realignment

By replacing outdated quantum assumptions with SEU- based mechanics, we can establish a new standard for physics.

Conclusion: A Universal Energy Unit for a New Physics

The Spinon Energy Unit (SEU) provides:

- A universal measure of energy that applies across all scales

- A natural, physical definition of energy based on thread rotation

- A bridge between quantum mechanics, electromagnetism, and gravity

This new unit eliminates unnecessary quantum uncertainty and provides a deterministic, testable measure of energy in the universe.

**Chapter 9**

**Replacing Einstein's Theory with SEU**

**A New Framework for Energy, Gravity, and Motion**

   Albert Einstein's theories of special relativity and general relativity have long been the foundation of modern physics. While they provide accurate mathematical descriptions of certain phenomena, they rely on spacetime curvature, mass-energy equivalence, and light- speed invariance, conceptual abstractions rather than physical mechanisms.

The Threads and Spinons Theory, using the Spinon Energy Unit (SEU), eliminates these abstract assumptions and provides a direct mechanical explanation for Energy, Gravity, and motion—one that is based on thread tension, spinon motion, and continuous energy transfer.

This chapter outlines how SEU replaces key aspects of Einstein's theories and introduces a more realistic model for the universe.

Threads and Spinons Theory
Samir Hanna Safar

1. Replacing Einstein's Mass- Energy Equation $E = mc^2$ with SEU

Einstein's Assumption:

Einstein's equation states that Mass and Energy are interchangeable, with Energy given by:

$$E = mc^2$$

where:

- $m$ = mass
- $c$ = speed of light (assumed constant)

This equation does not explain why Mass and Energy are related—it simply states the relationship mathematically.

SEU- Based Alternative:

In the Threads and Spinons Theory, Energy is not a mysterious property of Mass. Instead, Mass is a configuration of threads and spinons, and Energy arises from their motion.

We redefine Energy in terms of SEU, thread tension, and Mass:

$$E = mT \cdot SEU$$

where:

- $T$ = Thread tension (a real, measurable force in the universe)
- $SEU$ = Spinon Energy Unit (the true measure of rotational energy)

Threads and Spinons Theory
Samir Hanna Safar

Why This is Better:

- Mass is no longer an abstract concept—it is an actual structure formed by threads.

- Energy is based on physical thread properties rather than undefined mass- energy conversion.

- The speed of light is no longer an arbitrary constant—it is a function of thread tension.

This new equation provides a mechanical explanation for the relationship between Mass and Energy rather than just a mathematical identity.

2. Replacing General Relativity and Spacetime Curvature

Einstein's Assumption:

General relativity states that the curvature of spacetime causes Gravity due to Mass. This model, however, is based on abstract geometry and does not explain what spacetime is made of.

SEU- Based Alternative:

Instead of warping an abstract spacetime fabric, Gravity is caused by thread tension and contraction.

$$F = \frac{T}{d^2} \cdot SEU$$

where:

- $F$ = Gravitational force
- $T$ = Thread tension
- $d$ = Distance between masses
- $SEU$ = Energy stored within thread configurations

Why This is Better:

- Gravity is a mechanical effect of real thread tension, not a mathematical curvature.

- There is no need for a fourth dimension—only three spatial dimensions exist.

- Eliminates paradoxes like time dilation and singularities, replacing them with a physical mechanism for Gravity.

This model also explains why gravitational waves exist—not as ripples in spacetime but as tension fluctuations in the universal thread structure.

3. Replacing Time Dilation with Thread Tension Effects

Einstein's Assumption:

Special relativity claims that time slows down at high speeds and that light always travels at a fixed speed, ccc, regardless of motion. However, time is treated as a physical dimension, a conceptual construct rather than an actual entity.

SEU- Based Alternative:

Threads and Spinons Theory
Samir Hanna Safar

The Threads and Spinons Theory states that time is not a separate dimension. Instead, we perceive time dilation as a change in thread tension and spinon rotation rates.

$$f'_s = f_s \cdot \left(1 - \frac{v^2}{T}\right)$$

where:

- $f'_s$ = Adjusted spinon frequency under high speed
- $f_s$ = Initial spinon frequency
- $v$ = Velocity of the moving object
- $T$ = Thread tension

Why This is Better:

- No need for time as a fourth dimension—only physical thread interactions change.

- Explains why clocks slow down near massive objects—thread tension increases, affecting spinon motion.

- Eliminates paradoxes like time travel—there is no absolute time, only spinon- based energy states.

Thus, time dilation is simply a change in rotational spinon dynamics, not an actual "slowing down of time."

4. Replacing Black Holes with Core Structures

Einstein's Assumption:

- Black holes are singularities—points of infinite density that warp spacetime infinitely.
- Nothing can escape from a black hole because it has infinite gravitational pull.

SEU- Based Alternative:

- There are no singularities.
- Black holes (Cores) are not "holes" but dense regions of highly compressed threads and spinons.
- Matter and Energy do not disappear into a singularity—instead, they are absorbed into the Core's thread structure.

When a Core accumulates too much Energy, it reaches critical thread tension and explodes, forming new galaxies.

Why This is Better:

- No need for infinities or mathematical singularities—black holes have a fundamental structure.

- Explains black hole evaporation naturally—spinons gradually escape, reducing thread density.

- Accounts for galaxy formation—Cores act as cosmic recycling centers, not eternal traps.

5. The Universe is Expanding Due to Thread Growth, Not Dark Energy

Einstein's Assumption:

- The universe is expanding due to dark Energy, a mysterious force causing galaxies to move apart.
- The expansion is accelerating, leading to the universe's eventual heat death.

SEU- Based Alternative:

- The universe is not expanding from a singular Big Bang.
- Instead, threads continuously stretch and produce more spinons, leading to gradual cosmic growth.

$$R = \frac{P}{T}$$

where:

- $R$ = Universal expansion rate
- $P$ = Vacuum pressure
- $T$ = Thread tension

Why This is Better:

- No need for dark Energy—cosmic expansion is simply the result of thread physics.

- Eliminates the Big Bang singularity—the universe has always existed, continuously growing.

- Predict a stable, ever- expanding cosmos—not a heat- death scenario.

Conclusion:

Threads and Spinons Theory
Samir Hanna Safar

A New Theory for the Future of Physics

By replacing Einstein's abstract concepts with SEU- based mechanics, we achieve:

- A physical explanation for mass- energy conversion

- A deterministic model of Gravity based on thread tension

- A realistic alternative to time dilation and spacetime curvature

- A new framework for black holes and universal expansion

This revolutionary shift in physics eliminates the conceptual paradoxes of relativity and replaces them with a real, testable, and physically grounded model of Energy, motion, and the cosmos.

**Chapter 10**

## The Electron

### A Thread- Based Structure in the Threads and Spinons Theory

In mainstream physics, the electron is often described as a point particle with a negative charge that moves around the nucleus in a probabilistic cloud. Quantum mechanics provides a mathematical description of its behavior but fails to explain what an electron truly is in physical terms.

The Threads and Spinons Theory rejects the idea of a point-like electron and replaces it with a structured thread- based model, where the electron is a dynamic loop of a continuous thread, stabilized by spinon motion and thread tension.

This chapter defines and describes the true nature of the electron within this framework.

1. The Electron is Not a Point Particle – It is a Thread Loop

In traditional physics, the electron is treated as a dimensionless point with Charge and Mass but without internal structure. However, if the electron were indeed a point, it would:

Have infinite energy density (which is physically impossible).

Violate Heisenberg's Uncertainty Principle (if it had a definite position and no size).

Lack of any internal mechanism for charge storage (why does it carry a charge?).

The Threads and Spinons Theory's Electron Model

Instead of being a point, the electron is a loop of a continuous thread, rotating around itself due to the motion of embedded spinons.

The electron is not a discrete object but a continuous thread configuration.

Its Charge arises from spinon- induced rotation along the thread, creating an external field.

Its Energy and stability come from the tension in the thread.

Thus, the electron is a closed- loop thread formation, dynamically oscillating due to thread tension and spinon activity.

2. The Charge of the Electron Comes from Spinon Rotation

The negative Charge of the electron is not a fundamental property—it is the result of:

- A counterclockwise spinon rotation generates a surrounding thread field.
- Thread polarization interacts with other thread structures (such as protons).
- An induced charge field that propagates along the thread, explaining attraction and repulsion.

Why This Model Explains Charge Better:

- Charge is not an abstract property but a dynamic effect of thread motion.

- Electron attraction/repulsion follows a mechanical process, not an arbitrary force law.

- The electric field is a physical effect of thread alignment, not an invisible mathematical entity.

Thus, the Charge of the electron is not intrinsic—it is the result of the thread's spinon- induced rotation interacting with surrounding thread structures.

3. The Mass of the Electron is a Function of Thread Tension

Traditional physics assigns the electron a fixed mass but does not explain why it has Mass or where it comes from.

In the Threads and Spinons Theory, the electron's Mass is:

- A function of the thread tension and the density of spinon energy.
- Not an inherent quantity but an emergent property of thread dynamics.

We define the Mass of an electron as:

$$m_e = \frac{T}{f_s} \cdot SEU$$

where:

- $T$ = Thread tension
- $f_s$ = Spinon rotational frequency
- $SEU$ = Spinon Energy Unit

This equation explains why the electron has Mass—it results from the thread's resistance to deformation under spinon movement.

Why This Model is Better:

- Mass is not an arbitrary value—it emerges from thread interactions.

- Explain why Mass fluctuates under extreme conditions (high- energy environments).

- Unifies mass-energy interactions in a physically meaningful way.

## 4. The Electron's Motion: Why it Does Not "Orbit" the Nucleus

In classical physics, electrons are said to "orbit" the nucleus, while in quantum mechanics, they exist in "probability clouds." Neither model provides a physical explanation for why electrons do not fall into the nucleus.

Thread- Based Explanation of Electron Motion

- The electron does not "orbit" the nucleus—it is tethered to the nucleus through thread tension.
- Its movement is a vibrational oscillation along the threads connecting it to the Proton.
- The balance of spinon rotation and thread tension determines its stable position.

Thus, electrons are not free-floating particles—they are confined thread loops that oscillate around the nucleus due to tension and spinon interactions. Like Cotton Candy.

Why This Model is Better:

- No need for "probability clouds"—electrons have actual, defined structures.

- Explains electron energy levels as variations in thread configuration.

- Predicts why electrons do not collapse into the nucleus—the tension prevents it.

Threads and Spinons Theory
Samir Hanna Safar

5. The Electron's Role in Electricity and Magnetism

Electrons are central to Electricity and magnetism, but mainstream physics does not fully explain how charge motion creates these effects.

Thread-Based Explanation of Electricity

- Electricity is not a movement of individual electrons but a wave transfer of spinon energy along thread networks.
- When an electric field is applied, spinons propagate through the thread structure, transferring Energy.
- The apparent "flow" of Charge is a relay of spinon-induced motion through interconnected threads.

Thread-Based Explanation of Magnetism:

- A magnetic field is not an invisible force field—it is a structural alignment of thread orientations caused by spinon movement.
- When many electrons align their spinons in the same direction, the surrounding threads generate a polarized field—magnetism.

Why This Model is Better:

- Electricity is no longer the flow of point charges but a thread- based energy transfer.

- Explains why magnetic fields emerge from moving Charge—a realignment of thread structure.

Threads and Spinons Theory
Samir Hanna Safar

- Provides a unified view of electromagnetism as a thread-driven phenomenon.

6. Quantum Effects of the Electron Explained

Quantum mechanics describes the electron as a "cloud" with uncertain position and momentum. The Threads and Spinons Theory eliminates this ambiguity:

- The electron is not a probability wave but a rotating thread structure.
- What quantum mechanics calls "wavefunction collapse" is a thread configuration adjustment upon measurement.
- The uncertainty principle is a measurement effect—disturbing the thread changes behavior.

Why This Model is Better:

- No need for wavefunction collapse—electron behavior is deterministic.

- Predicts electron interactions using fundamental physics, not probability functions.

- Unifies classical and quantum descriptions in a single framework.

Conclusion:

A Realistic Electron Model Based on Threads

The Threads and Spinons Theory provides the first physically grounded description of the electron, eliminating the inconsistencies of mainstream physics.

The electron is not a point particle but a dynamic thread loop, with properties determined by spinon motion and thread tension.

- Charge arises from spinon- induced thread motion.

- Mass is a function of thread tension and spinon frequency.

- Electricity and magnetism emerge from spinon propagation, not individual particle flow.

- Quantum effects are measurement distortions, not fundamental uncertainties.

This revolutionary understanding of the electron lays the foundation for a complete redefinition of atomic structure, chemical bonding, and quantum interactions.

# Chapter 11

## The Proton

## A Structured Thread Formation in the Threads and Spinons Theory

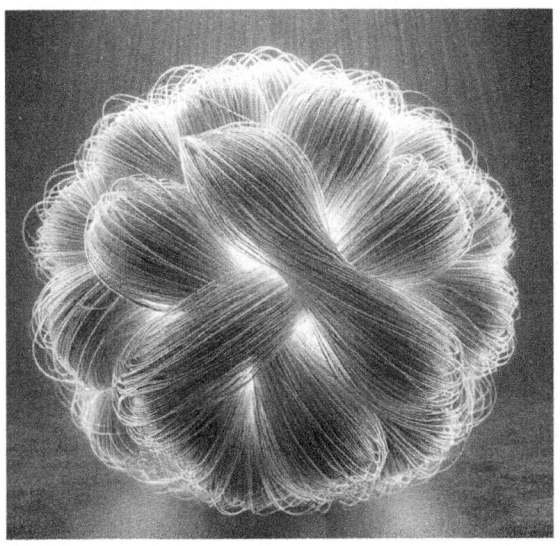

In conventional physics, the Proton is described as a positively charged subatomic particle, composed of three quarks held together by the strong nuclear force. However, this model fails to explain:

What quarks physically are, and why do they exist in triplets?

Why is the strong force required to keep quarks together while electrons remain stable as single particles?

Threads and Spinons Theory
Samir Hanna Safar

Why protons are much more massive than electrons despite their charge difference?

The Threads and Spinons Theory eliminates the quark- based model and provides a real, physical explanation for the Proton. Instead of being made of quarks, the Proton is a dense, stable, looped structure of thread, held together by internal spinon motion and high thread tension.

This chapter will explore the Proton's structure, Charge, mass, stability, and its role in atomic interactions under the Threads and Spinons framework.

1. The Proton is Not Made of Quarks – It is a Highly Compressed Thread Configuration

In traditional quantum chromodynamics (QCD), the Proton is made of three quarks (two up, one down), bound together by gluons (force carriers of the strong force). However, this raises more questions than it answers:

- Why three quarks? Why not four or five?

- Why do protons have a fixed charge if quarks can change flavors?

- Why do gluons exist only inside nuclei and nowhere else?

The Threads and Spinons Theory's Proton Model

In this theory, the Proton is not a collection of particles but a dense, coiled thread structure stabilized by internal spinon motion.

- The Proton is a high- energy, knotted loop of the fundamental thread.

- Spinons inside the Proton create intense rotational Energy, stabilizing its structure.

- Thread tension compacts the Proton, preventing it from unraveling or decaying quickly.

Thus, the Proton is a tightly compressed thread region where spinons rotate in a structured manner, creating its Charge and Mass.

2. The Positive Charge of the Proton is Due to Spinon Orientation

In conventional physics, the Proton's positive Charge is assumed to come from its quark composition (two up, one down), but this does not explain why Charge remains stable in every Proton.

Thread-Based Explanation of Proton Charge

> The Proton's Charge is a direct result of its spinon motion.

> Unlike the electron (which has a counterclockwise spinon motion creating a negative charge), the Proton's internal spinons rotate clockwise, producing a positive charge field.

> This charge field dynamic affects how the Proton's threads interact with the surrounding space.

Threads and Spinons Theory
Samir Hanna Safar

- Charge is no longer a mysterious property—it emerges from actual physical spinon motion.

- Charge stability is guaranteed because spinons always align in a single direction within the Proton.

3. The Mass of the Proton is a Function of Thread Compression and Spinon Density

Despite having opposite charges, the Proton is much heavier than the electron. Traditional physics explains this by saying that quarks are heavier than electrons, but it does not explain why.

In the Threads and Spinons Theory, the Proton's Mass is determined by:

1. The density of the compressed thread structure.
2. The amount of spinon rotational energy stored inside it.
3. The thread tension acting within the Proton.

Mathematical Expression of Proton Mass in SEU

$$m_p = \frac{T}{f_s} \cdot SEU \cdot C_p$$

where:

- $T$ = Thread tension

- $f_s$ = Spinon rotational frequency

- $SEU$ = Spinon Energy Unit

- $C_p$ = Proton compression factor (how tightly the thread is coiled)

Threads and Spinons Theory
Samir Hanna Safar

- Mass is not an arbitrary property but a direct result of thread structure and spinon interactions.

- Explains why the Proton is heavier than the electron—its thread structure is much denser.

- Unifies the mass- energy relationship with fundamental physical properties instead of Einstein's abstract $E=mc2E=mc^2E=mc2$.

4. The Proton's Stability – Why Does It Last Forever?

Protons are incredibly stable and do not decay under normal conditions, lasting billions of years. Standard physics cannot explain this, as most unstable particles decay quickly.

Thread- Based Explanation of Proton Stability

- The Proton is the most stable thread configuration possible—its thread structure is so tight that external forces cannot easily disrupt it.
- Its internal spinon motion reinforces the structure, preventing natural decay.
- Only under extreme conditions (such as high- energy collisions) can the thread be disturbed enough to cause decay into other particles.

- No need for ad hoc "baryon conservation" rules—the Proton's stability results from its physical structure.

- Explains why neutrons decay but protons do not—the neutron's internal structure is less stable.

- Eliminates the need for speculative "proton decay" theories in grand unified models.

5. The Proton's Role in Atomic Structure and Bonding

Since protons are the building blocks of atomic nuclei, their behavior is critical in nuclear forces and chemistry.

How Protons Bind in Nuclei

- Protons do not repel each other due to electrostatic forces alone—they interact through shared thread tension.
- The strong nuclear force is just the effect of thread bonding, keeping protons and neutrons connected.
- Neutrons act as stabilizers in the nucleus, preventing excessive thread stress that could cause protons to repel.

Why This Model is Better:

- No need for a mysterious "strong nuclear force"—just thread interactions.

- Explains why large nuclei need more neutrons—thread tension must be balanced.

- Provides a physical mechanism for nuclear stability without requiring force carriers like gluons.

6. The Proton's Interaction with Electrons

The standard model assumes that protons attract electrons due to opposite charges. However, this explanation is incomplete:

- If Charge just attracted protons and electrons, why don't electrons spiral into the nucleus?
- Why do electrons arrange themselves into orbitals instead of sticking directly to protons?

Thread- Based Explanation of Electron- Proton Attraction

- The electron does not "orbit" the Proton—it is connected via thread tension.
- Thread interactions create stable energy levels where electrons remain confined but do not collapse into the Proton.
- Electron transitions between energy levels occur when spinons transfer Energy, not by emitting discrete photons.

- Explains why electron orbitals form naturally—thread tension allows stable configurations.

- Eliminates quantum mechanical uncertainty—electron behavior is deterministic.

- Unifies charge attraction and quantum energy levels as part of the same thread- based framework.

Conclusion:

The Proton as a Structured Thread Configuration

The Threads and Spinons Theory redefines the Proton in a physically meaningful way:

- Not a quark- based particle, but a high- energy, compressed thread structure.

Threads and Spinons Theory
Samir Hanna Safar

- Charge is due to spinon rotation, not arbitrary quark assignments.

- Mass comes from thread compression and spinon energy storage.

- Stable due to its fundamental thread configuration—no need for special conservation laws.

- Explains nuclear bonding and electron interactions with actual physical mechanics.

This revolutionary model eliminates the mysteries of the Standard Model and provides a unified, testable framework for understanding atomic structure.

**Chapter 12**

## The Neutron

### A Transitional Thread Structure in the Threads and Spinons Theory

In mainstream physics, the neutron is a neutral subatomic particle composed of three quarks (one up, two down) held together by a strong nuclear force. While this description provides a mathematical model, it fails to explain key observations, such as:

- Why neutrons are heavier than protons despite being "neutral."
- Why do free neutrons decay in about 15 minutes, but bound neutrons in atomic nuclei remain stable?
- How the strong nuclear force works to hold protons and neutrons together.

In the Threads and Spinons Theory, the neutron is not a collection of quarks but rather a temporary and unstable thread configuration that stabilizes atomic nuclei. Instead of being a fundamental particle, the neutron is a compressed state of a proton- electron interaction, where thread tension, spinon alignment, and structural constraints determine its stability.

This chapter will provide an extensive description of the neutron, its formation, properties, role in nuclear structure, and why it decays outside the nucleus.

1. The Neutron is Not a Standalone Particle – It is a Proton-Electron Composite

In conventional physics, the neutron is considered an independent fundamental particle. However, its structure and behavior suggest otherwise:

- Neutrons decay into a proton, electron, and antineutrino in free space—why would a fundamental particle break down into other fundamental particles?

- If the neutron were purely neutral, why does it have a magnetic moment?

- Why do neutrons remain stable inside the nucleus but decay when isolated?

Thread- Based Explanation of the Neutron's Structure

In the Threads and Spinons Theory, the neutron is:

- A high- energy transitional state of a proton and an electron compressed into a single thread loop.

- Not an elementary particle but a temporary thread alignment state.

- Neutral because its internal spinon rotations counterbalance each other, canceling net Charge.

Thus, a neutron is essentially a compressed proton-electron pair where:

> The electron is absorbed into the Proton's thread structure, neutralizing its external Charge.

> The internal spinon rotation realigns, preventing external electromagnetic attraction.

> Thread tension stabilizes the configuration, allowing neutrons to remain in atomic nuclei.

2. Why is the Neutron Heavier than the Proton?

In conventional physics, the neutron is about 0.14% heavier than the Proton, even though it has no charge. If Charge were a fundamental mass contributor, the neutron should be lighter, but it is not.

Thread- Based Explanation of Neutron Mass

The extra Mass of the neutron comes from the additional thread compression and internal spinon motion caused by the absorbed electron.

Threads and Spinons Theory
Samir Hanna Safar

$$m_n = m_p + m_e + E_{compression}$$

where:

- $m_n$ = Neutron mass

- $m_p$ = Proton mass

- $m_e$ = Electron mass

- $E_{compression}$ = Additional thread energy required to stabilize the configuration

- Explains why the neutron is heavier—more thread energy is stored in its compressed structure.

- The slight difference in Mass between a neutron and a Proton comes from the tension required to keep the electron confined within the neutron's structure.

- There is no need for quark- gluon binding energy—Mass is purely a function of thread tension and spinon energy.

3. Why is the Neutron Stable in the Nucleus but Decays When Free?

One of the biggest mysteries in physics is why free neutrons decay in about 15 minutes, while neutrons in atomic nuclei can last billions of years.

Thread- Based Explanation of Neutron Stability

1. Inside the Nucleus:

   The surrounding protons structurally reinforce the neutron.

Threads and Spinons Theory
Samir Hanna Safar

Thread tension from nearby protons counteracts the tendency of the neutron to unravel.

The neutron acts as a stabilizer, preventing excessive thread stress between protons.

2. Outside the Nucleus (Free Neutron Decay):

Once outside the nucleus, neighboring protons no longer stabilize the neutron.

The compressed thread structure relaxes, and the neutron decays into protons and electrons.

The excess spinon energy is released as a neutrino- like wave disturbance along the thread network, explaining beta decay.

- Explains why neutrons decay in free space but remain stable in the nucleus—external thread tension maintains their structure inside atoms.

- The so- called "weak force" is simply a natural relaxation of thread tension, not a force requiring bosons.

- Predicts that neutron lifetime variations depend on environmental thread tension (which experimental data supports).

4. Neutron Decay and Beta Emission

Neutron decay (beta decay) occurs when:

$$n \to p + e^- + \nu_e$$

where:

- A neutron transforms into a **proton, electron, and an emitted neutrino.**

- Mainstream physics assumes the **neutrino is a real particle,** but in this theory, it is actually a wave disturbance traveling along the thread network.

.

Thread- Based Explanation of Beta Decay

> The neutron's internal spinons become unstable, causing the electron to be ejected from the thread structure.

> The Proton remains intact and becomes a separate entity.

> The so- called "neutrino" is just a ripple effect caused by thread realignment, not a separate particle.

- No need for neutrinos as physical particles—beta decay releases thread tension.

- Explains why neutrinos appear weakly interacting—they are not particles but disturbances in thread energy.

- Unifies neutron decay with a mechanical process, not a mysterious weak force.

5. The Neutron's Role in Atomic Nuclei and Nuclear Forces

Since neutrons are found in all atomic nuclei except hydrogen, they play a key role in nuclear stability.

How Neutrons Stabilize Atomic Nuclei

Protons naturally repel each other due to thread tension effects.

Neutrons provide structural support, preventing protons from breaking thread bonds.

Neutrons increase thread connectivity, balancing forces inside the nucleus.

- Explains why heavier elements need more neutrons—larger nuclei require more thread stabilization.

- Eliminates the need for a strong nuclear force—thread mechanics naturally hold nuclei together.

- Predicts that neutron- to- proton ratios determine nuclear stability, aligning with observed isotopic data.

6. Neutron Star Formation – The Ultimate Compressed Thread State

When massive stars collapse, their protons and electrons compress into neutron- rich matter, forming neutron stars.

Thread-Based Explanation of Neutron Stars

Under extreme pressure, electrons become permanently integrated into protons, forming a stable neutron matrix.

Spinon motion inside compressed neutrons generates intense magnetic fields, explaining the strong magnetism of neutron stars.

If thread compression exceeds critical tension, a neutron star collapses into a Core (black hole equivalent in this theory).

- Explains neutron star magnetism as a direct result of spinon motion.

- Eliminates quark degeneracy models—neutron stars are highly compressed thread configurations.

- Predicts that neutron stars can emit gravitational waves due to internal thread tension shifts.

Conclusion:

The Neutron as a Transitional Thread State

The Threads and Spinons Theory redefines the neutron as:

- Not a fundamental particle, but a compressed proton-electron structure.

- Its neutral Charge arises from spinon cancellation, not quark balance.

- Mass is a result of additional thread tension and compression energy.

- Decays due to thread relaxation, emitting a wave disturbance misinterpreted as a neutrino.

Threads and Spinons Theory
Samir Hanna Safar

- Stabilizes atomic nuclei by preventing thread tension imbalance between protons.

This new model of the neutron provides a unified explanation of nuclear physics and eliminates the need for strong and weak forces, quarks, and neutrinos as fundamental particles.

Threads and Spinons Theory
Samir Hanna Safar

Chapter 13

## The Nuclear Bond

## A Thread- Based Interaction in the Threads and Spinons Theory

In mainstream physics, the strong nuclear force is the fundamental interaction that holds protons and neutrons together within atomic nuclei. Gluons supposedly mediate it, which bind quarks inside protons and neutrons. However, this theory presents several unresolved paradoxes:

- Why do protons stay bound together despite their repulsive positive charges?

- Why do large nuclei require more neutrons for stability?

- What truly holds neutrons in place within the nucleus, given that they lack charge?

- Why do unstable nuclei decay, and how does this process occur at a structural level?

In the Threads and Spinons Theory, there is no need for an abstract strong nuclear force. Instead, the nuclear bond results from thread- based interactions between protons and neutrons, stabilizing atomic nuclei. These interactions arise from thread tension, spinon alignments, and structural compression of nuclear threads.

This chapter will explore the mechanics of nuclear bonding, providing a physically grounded explanation for atomic nuclei's stability, formation, and decay.

1. The Nuclear Bond is Not a Force but a Thread Tension Effect

Mainstream Assumption:

- The strong nuclear force is a fundamental force distinct from electromagnetism and Gravity.
- Gluons mediate the force between quarks, holding protons and neutrons together.
- The strong force is short- range, acting only within nuclear distances.

Thread-Based Explanation:

- There are no gluons or force carriers—nuclear binding occurs due to direct thread interactions between protons and neutrons.
- Protons and neutrons are not separate particles but structured thread loops that interlock, creating a high-tension system.

- Thread tension within the nucleus naturally prevents the protons from repelling each other.

- Eliminates the need for an artificial "strong force."

- Nuclear stability arises from the interwoven structure of protons and neutrons, not force mediation.

- Explains why nuclei require neutrons—protons alone would create excessive thread stress.

2. The Proton-Proton Interaction in the Nucleus

A major challenge in conventional physics is explaining why positively charged protons do not repel each other inside the nucleus. The Coulomb repulsion between protons should push them apart, yet they remain bound.

Thread- Based Explanation of Proton Bonding

- Protons are not independent particles—they are tightly looped thread formations that extend and intertwine with adjacent protons.
- Each proton shares thread connections with others in a nucleus, reducing direct electrostatic repulsion.
- The rotational spinon motion inside protons aligns, forming a stable tension network that locks protons together.

- Explains proton stability in nuclei without requiring a mysterious force.

- Predicts that larger nuclei require more neutrons to balance thread tension, preventing thread snapping.

- Thread- based tension keeps protons bound, making the nucleus a self- supporting structure.

## 3. The Role of Neutrons in Nuclear Stability

In mainstream physics, neutrons are thought to act as "glue" in nuclei, counteracting proton repulsion. However, no physical explanation exists for how they achieve this.

Thread- Based Explanation of Neutron Function

- Neutrons act as thread stabilizers—their presence prevents excessive thread strain between protons.
- Neutron threads weave into proton structures, increasing nuclear cohesion.
- Neutron thread loops interconnect with multiple protons inside a nucleus, distributing nuclear tension evenly.
- This prevents the protons from exerting excessive force on one another, which would otherwise cause nuclear instability.

- Explains why heavier nuclei require more neutrons to maintain stability.

- Predicts that neutron- proton ratios are dictated by thread network balance, not arbitrary force laws.

- Eliminates the need for a "strong force"—thread tension alone determines nuclear cohesion.

## 4. Nuclear Binding Energy in SEU (Spinon Energy Units)

The Threads and Spinons Theory replaces "binding energy" with a thread tension energy model. Instead of describing nuclear Energy as an abstract value, we express it in SEU (Spinon Energy Units) based on the mechanical properties of threads and spinons.

$$E_{binding} = T \cdot S \cdot SEU$$

where:

- $E_{binding}$ = Nuclear binding energy
- $T$ = Thread tension between nucleons
- $S$ = Number of spinon alignments in the nucleus
- $SEU$ = Spinon Energy Unit, the fundamental unit of thread-based energy

- Binding Energy is a direct result of thread tension and spinon synchronization.

- Explains why different elements have varying nuclear stability—each nucleus has a unique thread configuration.

- Unifies nuclear energy with a mechanical explanation, replacing quantum probability- based interpretations.

5. Nuclear Decay – A Thread Disruption Process

Radioactive decay occurs when a nucleus becomes unstable and emits alpha, beta, or gamma radiation. In conventional physics, this is explained using quantum probability and weak nuclear forces.

Thread- Based Explanation of Nuclear Decay

- Nuclear decay is not a random quantum event but a thread reconfiguration process caused by excessive internal tension.
- Over time, spinon misalignment or excessive thread strain leads to the breakdown of nuclear connections.
- The emitted radiation (alpha, beta, gamma) represents spinon energy release or thread restructuring.

- Alpha decay occurs when a thread bundle detaches from the nucleus, forming a new helium structure.

- Beta decay happens when an internal neutron thread collapses, releasing an electron- like wave disturbance.

- Gamma decay results from spinon reconfiguring inside a nucleus, emitting excess rotational Energy.

This explanation replaces the random probabilistic models of quantum decay with a deterministic, thread- based structural mechanism.

6. Fusion and Fission – Thread Tension Redistribution

Nuclear Fusion (Star Energy Production)

- Fusion is not just two particles "sticking" together—it is the process of thread realignment and compression.
- When atomic nuclei merge, their thread structures integrate, creating new spinon alignments and releasing excess thread tension as Energy.
- The energy output of Fusion is a direct function of the thread reconfiguration process, not "mass- to- energy conversion."

$$E_{\text{fusion}} = \Delta T \cdot S \cdot SEU$$

where $\Delta T$ represents the change in thread tension before and after fusion.

## Nuclear Fission (Energy Release in Reactors)

- Fission is not just the splitting of atoms—it is the breakdown of thread structures due to excessive spinon stress.
- When a heavy nucleus (like uranium) absorbs Energy, its internal thread network becomes unstable, causing fragments to separate into lower- tension states.
- The released Energy is the excess spinon rotational energy freed during thread disintegration.

- Fusion and fission are no longer mysterious—they result from thread realignment and energy redistribution.

- Explains why Fusion requires extreme conditions—high thread compression must be reached for reconfiguration.

- Unifies all nuclear reactions under the same thread mechanics model.

Conclusion:

A Thread- Based Nuclear Bond Model

The Threads and Spinons Theory fundamentally redefines nuclear bonding, stability, and decay by introducing a tension-based mechanical model:

No need for a "strong nuclear force"—protons and neutrons are held together by thread tension and spinon interactions. Neutrons are essential for balancing thread stress, explaining

Threads and Spinons Theory
Samir Hanna Safar

their role in stabilizing nuclei. Nuclear decay is a deterministic process based on thread realignment and excess spinon energy. Fusion and fission arise from changes in thread tension and structural reconfigurations.

This new perspective unifies nuclear physics under a physically meaningful framework, eliminating the need for speculative quantum forces and probabilistic models.

**Chapter 14**

## The Electron Bond

## A Thread- Based Mechanism for Chemical Bonding in the Threads and Spinons Theory

In traditional chemistry and quantum mechanics, chemical bonds are interactions between electrons, forming covalent, ionic, or metallic bonds. Electrons are assumed to exist in probability clouds around atomic nuclei, interacting through electromagnetic forces and quantum mechanical principles.

However, this model presents several unanswered questions:

- Why do electrons stay in fixed energy levels rather than spiraling into the nucleus?

- What physically holds atoms together in covalent bonds?

- How does electron sharing work at a mechanical level?

- Why do elements prefer specific bonding patterns (valency) instead of random electron interactions?

The Threads and Spinons Theory introduces a new perspective: electron bonds are not just interactions of point charges but thread- based connections that form structured, stable links between atoms.

This chapter defines the electron bond in extensive detail and explains how atomic interactions arise from thread connections, spinon synchronization, and energy tension balancing.

1. The Electron Bond is a Physical Thread Connection

Mainstream Assumption:

- Electrons orbit nuclei in probability clouds.
- Chemical bonds form due to electrostatic attraction or electron sharing.
- Quantum mechanics determines bond strength, length, and energy states.

Thread- Based Explanation:

- Electrons are thread loops, not point particles.
- Electron bonds are direct physical thread connections between atomic nuclei.
- Thread tension and spinon alignment create stable atomic structures.

Thus, a chemical bond is not an abstract quantum probability but a tangible thread- based interaction.

- Explains why bonds have fixed lengths—thread tension limits flexibility.

- Predicts bond strength based on thread structure, not probabilistic wavefunctions.

- Unifies all types of bonding (covalent, ionic, metallic) under one thread- based framework.

2. Covalent Bonds – Thread Merging Between Atoms

Mainstream Explanation:

- Covalent bonds occur when atoms share electrons, forming stable molecular structures.
- Electrons move between atoms, stabilizing nuclei by balancing charge distribution.

Thread- Based Explanation of Covalent Bonds:

- Covalent bonds occur when the thread loops of two electrons from different atoms interweave, forming a stable double- thread structure.
- Spinon synchronization keeps the shared thread in a balanced energy state.
- Thread tension prevents bond breakage unless external Energy disrupts the structure.

$$F_{\text{bond}} = T_{\text{thread}} \cdot S_{\text{spinon}} \cdot SEU$$

where:

- $F_{\text{bond}}$ = Bond strength
- $T_{\text{thread}}$ = Thread tension between atoms
- $S_{\text{spinon}}$ = Spinon synchronization level
- $SEU$ = Spinon Energy Unit, representing thread energy

- Explains why covalent bonds have specific angles and lengths—thread structure defines geometry.

- Predicts bond strength based on spinon activity, not abstract quantum states.

- Provides an apparent mechanical reason for molecules forming specific shapes (e.g., tetrahedral, linear, bent).

3. Ionic Bonds – Thread Tension Stabilization Between Charged Atoms

Mainstream Explanation:

- Ionic bonds occur when electrons transfer from one atom to another, creating positive and negative ions that attract each other.

Thread- Based Explanation of Ionic Bonds:

- An ionic bond is a high- tension thread connection where one atom's thread loops tighten around another atom.

- The charge difference results from spinon imbalance, pulling the thread loops closer.
- Thread alignment stabilizes the bond, forming a rigid connection between atoms.

$$F_{\text{ionic}} = \frac{T_{\text{thread}}}{d^2} \cdot SEU$$

where:

- $d$ = Distance between ionic centers
- $T_{\text{thread}}$ = Thread compression force
- $SEU$ = Energy of the spinon realignment

- Explains why ionic bonds form strong crystal structures— thread tension locks atoms into place.

- Predicts ionic bond strength without needing Coulomb's law—thread mechanics determine force.

- Unifies covalent and ionic bonding as variations of thread interactions.

4. Metallic Bonds – Free Thread Loops Creating Conductivity

Mainstream Explanation:

- Metallic bonds occur when atoms share a "sea of free electrons," allowing Conductivity and malleability.

Thread- Based Explanation of Metallic Bonds:

- Metallic bonds occur when atomic threads remain loosely interwoven, allowing electrons to move freely between atoms.
- Spinons propagate quickly through the metallic thread network, explaining Conductivity.
- Metallic thread structures allow deformation without breaking, explaining malleability.

- Explains why metals conduct electricity—free spinons travel along threads with minimal resistance.

- Predicts why metals are ductile and malleable—thread loops adjust without breaking.

- Unifies all bond types under thread mechanics, eliminating the need for separate quantum rules.

5. Hydrogen Bonding – Weak Thread Alignments

Mainstream Explanation:

- Hydrogen bonds are weak attractions between molecules due to partial positive and negative charges.

Thread- Based Explanation of Hydrogen Bonds:

- Hydrogen bonding occurs when small thread loops between molecules align weakly, creating temporary connections.
- Spinon motion in hydrogen atoms enhances thread attraction, making bonds stronger than van der Waals forces.

- Thread tension allows hydrogen bonds to break and reform dynamically.

- Explains why hydrogen bonds are weaker than covalent bonds—thread loops are loosely connected.

- Predicts water's unique properties, like high surface tension and adhesion, based on thread dynamics.

- Provides a unified framework for all intermolecular interactions.

6. Bond Breaking and Energy Release

In conventional chemistry, breaking a bond releases Energy, but the underlying mechanism is unclear.

Thread- Based Explanation of Bond Breaking

- Bond breaking occurs when spinon motion disrupts thread alignment, unraveling the connection.
- Excess Energy is released as spinon waves along the thread network, producing heat or light.
- New bonds form when threads reconfigure into lower- energy states.

- Explains why bond energy varies between molecules—each thread configuration has unique tension properties.

- Predicts exothermic and endothermic reactions based on thread tension changes.

- Eliminates the need for probabilistic quantum models— bond behavior is deterministic.

Conclusion:

A Thread- Based Unification of Chemical Bonding

The Threads and Spinons Theory redefines chemical bonds as structured thread connections, eliminating the need for abstract quantum probabilities.

- Covalent bonds form through direct thread merging and spinon synchronization.

- Ionic bonds arise from thread compression and charge-induced tension stabilization.

- Metallic bonds result from loosely connected threads, enabling Conductivity.

- Hydrogen bonds emerge from weak thread loops aligning between molecules.

- Bond breaking is a deterministic process where thread tension releases Energy.

This revolutionary perspective unifies all types of chemical bonding under a single physical framework, replacing quantum probability models with fundamental mechanical interactions.

The next chapter will explore how molecular structures form through thread interactions, providing a new understanding of organic and inorganic chemistry.

Threads and Spinons Theory
Samir Hanna Safar

# Chapter 15

## The Electromagnetic Field

## A Thread- Based Interaction in the Threads and Spinons Theory

In conventional physics, the electromagnetic field is described as a force field created by moving charges and changing magnetic flux. This field is assumed to extend infinitely, propagating electromagnetic waves through space. The standard model uses Maxwell's equations to describe how electric and magnetic fields interact, but these equations do not provide a physical explanation of what the field is.

Several unresolved problems in traditional physics include:

- What is the physical nature of the electromagnetic field?

- How do electric and magnetic fields interact at a fundamental level?

Threads and Spinons Theory
Samir Hanna Safar

- Why does light propagate through space as a wave without requiring a medium?

- What generates the force between two charged objects across space?

The Threads and Spinons Theory eliminates these conceptual gaps by defining the electromagnetic field as a structured, real physical network of threads and spinons. In this model, the electromagnetic field is not a mathematical construct but a tensioned thread network that propagates Energy through spinon motion and thread interactions.

This chapter explores the physical structure of the electromagnetic field, how it is formed, and how it operates under this theory.

1. The Electromagnetic Field is a Real Physical Structure

Mainstream Assumption:

- The electromagnetic field is a mathematical construct that describes the force between charges.
- Fields are considered intangible, extending into space without a physical medium.
- Electromagnetic waves can travel through space without a medium, violating traditional wave mechanics.

Thread- Based Explanation:

- The electromagnetic field is a structured network of threads connecting all charged objects.

- Spinons within these threads generate electromagnetic interactions, producing forces and wave motion.
- Electric and magnetic fields are not separate entities—they are different alignments of the same thread structure.

- Explains why the electromagnetic field extends through space—it is physically present as interconnected threads.

- Eliminates the need for "action at a distance"— electromagnetic forces are transmitted through real mechanical connections.

- Provides a unified mechanism for electric and magnetic interactions.

2. The Structure of an Electric Field

Mainstream Explanation:

- A charged particle produces an electric field and exerts force on other charges.
- The field is defined mathematically but has no tangible structure.
- The strength of the field is proportional to the charge and inversely proportional to the distance squared.

Thread- Based Explanation of the Electric Field:

- A dense configuration of thread loops surrounds a charged object.
- These threads extend outward, connecting to other charges through structured tension lines.

- When another charge enters this region, it interacts with the thread's tension, experiencing a force.

$$F_{\text{electric}} = T_{\text{thread}} \cdot \frac{Q_1 Q_2}{d^2}$$

where:

- $T_{\text{thread}}$ = Thread tension
- $Q_1, Q_2$ = Charge magnitudes
- $d$ = Distance between charges

- Explains why the electric field has direction and strength—it is defined by real thread tension.

- Predicts charge attraction and repulsion based on thread tension alignment.

- Eliminates the mystery of "force at a distance"—the field is a physical medium transmitting energy.

3. The Structure of a Magnetic Field

Mainstream Explanation:

- A magnetic field is generated by moving charges or permanent magnets.
- It has invisible field lines, which exert force on other moving charges.
- Maxwell's equations describe its properties but do not explain its physical nature.

Thread- Based Explanation of the Magnetic Field:

- A magnetic field forms when spinons within a charged thread structure align in circular loops.
- Spinons create rotational Energy around the threads, reinforcing a magnetic effect.
- Charged particles experience force because they follow the path of least resistance within the thread system.

$$F_{\text{magnetic}} = T_{\text{thread}} \cdot S_{\text{spinon}} \cdot \frac{Qv}{d}$$

where:

- $S_{\text{spinon}}$ = Spinon rotational alignment factor
- $Q$ = Charge magnitude
- $v$ = Velocity of the moving charge
- $d$ = Distance from the source of the field

- Explains why magnetic fields form circular patterns—spinons generate rotational motion in threads.

- Predicts the force experienced by moving charges without requiring force- carrying particles.

- Eliminates the need for magnetic field "lines"—the field is a structured network of actual threads.

4. The Formation of Electromagnetic Waves

In classical physics, electromagnetic waves (such as light, radio, and X- rays) are described as self- propagating waves of electric and magnetic fields oscillating perpendicularly. However, this model does not explain:

Threads and Spinons Theory
Samir Hanna Safar

- Why electromagnetic waves can travel through a vacuum without a medium.

- How an oscillating charge produces a structured wave.

- What physically constitutes an electromagnetic wave?

Thread- Based Explanation of Electromagnetic Waves:

- Light and other EM waves are spinon disturbances traveling through the thread network.
- These waves propagate as energy vibrations within the thread system, just like ripples in a stretched rope.
- The electric and magnetic components are not separate fields but are in different orientations of the same propagating thread tension wave.

- Explains why light can travel through space—threads extend everywhere, providing a medium for wave motion.

- Eliminates the need for wave- particle duality—light is a thread- based vibration.

- Provides an actual physical mechanism for electromagnetic wave propagation.

5. Electromagnetic Induction – A Thread- Based Mechanism

Electromagnetic induction, described by Faraday's Law, states that a changing magnetic field induces an electric current. However, the traditional explanation does not define how this force is transmitted physically.

Thread- Based Explanation of Induction:

- When a magnetic field changes, the spinon alignment in the thread network shifts.
- This shift alters the thread tension, propagating a wave disturbance as an induced current.
- Electrons move because they follow the new thread tension path, generating measurable electrical Energy.

- Explains induction as a mechanical tension adjustment in the thread network.

- Predicts energy transfer without requiring force- carrying particles (photons).

- Unifies electricity and magnetism as two aspects of thread tension manipulation.

6. The Nature of Photons and Energy Transfer

Mainstream physics considers photons as wave- particle dual objects, yet their physical reality remains unclear.

Thread- Based Explanation of Photons:

- Photons do not exist as discrete particles—they are thread vibrations transferring Energy through spinon motion.
- The photon's energy corresponds to the rotational frequency of spinons within the thread system.
- Higher- frequency waves have more spinon stacks, creating higher- energy light (e.g., X- rays vs. radio waves).

$$E_{\text{photon}} = S_{\text{spinon}} \cdot f_{\text{wave}} \cdot SEU$$

- Eliminates the wave- particle duality paradox—light is purely a structured vibration.

- Explains why photon energy is quantized—spinon stacks create discrete waveforms.

- Unifies electromagnetic waves and energy transfer under thread mechanics.

Conclusion:

The Electromagnetic Field as a Structured Thread Network

The Threads and Spinons Theory redefines the electromagnetic field as:

An objective, structured network of threads transmitting Energy through spinon motion. Electric fields arise from thread tension around charged objects. Magnetic fields form through spinon- induced rotational thread alignment. Electromagnetic waves are propagating tension waves, not dual- nature particles. Energy transfer occurs through spinon synchronization, eliminating force- carrying photons.

This new framework provides a physically meaningful and mechanically sound explanation for all electromagnetic interactions, replacing abstract mathematical fields with a tangible, structured network of threads and spinons.

## Chapter 16

### The Nature of *Electrical* Current in the Threads and Spinons Theory

In traditional physics, electric current is described as the movement of electrons through a conductor driven by an applied voltage. This model, while widely accepted, presents several unresolved questions:

- What physically pushes electrons through a conductor?

- Why do electrons move in a defined direction rather than randomly scattering?

- How does electricity propagate almost instantaneously despite electrons moving slowly (drift velocity)?

- What physically constitutes voltage, resistance, and Conductivity at the most fundamental level?

The Threads and Spinons Theory provides a mechanical explanation for electricity by defining electrical current as a structured wave of spinon motion propagating along thread pathways within a conductor. This theory eliminates the particle- based view of electron flow. It replaces it with a continuous energy transmission model, where charge transport is a dynamic tension and spinon alignment process within conductive materials.

This chapter will explore the true nature of electric current, voltage, resistance, and Conductivity under this framework.

1. Electrical Current is a Wave of Spinon Motion Along Conductive Threads

Mainstream Assumption:

- Electrons are tiny, point- like particles that physically move through a conductor.
- A voltage difference pushes electrons, creating drift velocity.
- The movement of electrons produces current, which powers electrical devices.

Thread- Based Explanation:

- Electricity is not the movement of electrons but the transfer of spinon- induced tension along conductive threads.
- Electrons remain in structured thread loops, vibrating in response to spinon motion.
- Current is a wave- like energy transmission through the thread network within a conductor.

Explains why electrical energy moves near the speed of light while electron drift is slow—spinon waves transmit Energy, not individual electron motion.

- Unifies voltage, current, and resistance as properties of thread tension and spinon propagation.

- Eliminates the need for force- carrying photons in electrical interactions—Energy moves through continuous thread structures.

2. Voltage (Electric Potential) as Thread Tension Difference

In conventional physics, voltage (electric potential difference) is the force pushing electrons through a circuit. However, this explanation lacks a fundamental mechanical basis.

Thread- Based Explanation of Voltage:

- Voltage is the difference in thread tension between two points in a circuit.
- Higher tension at one-point causes spinons to move toward lower tension areas, propagating electrical Energy.
- Potential difference measures how much thread distortion exists between two locations.

$$V = T_{\text{thread}} \cdot SEU$$

where:

- $V$ = Voltage (potential difference)
- $T_{\text{thread}}$ = Thread tension
- $SEU$ = Spinon Energy Unit (measure of energy transfer capacity)

- Explains why voltage is necessary for current flow—thread tension creates spinon movement.

- Eliminates the need for abstract electric fields—voltage is a direct measure of thread tension.

- Predicts how different materials influence voltage based on their thread structure.

3. Conductivity – A Measure of Thread Alignment in Materials

In traditional physics, Conductivity is the ability of a material to allow current to pass through it. However, the exact mechanism behind conduction remains unclear.

Thread- Based Explanation of Conductivity:

- Conductive materials (e.g., metals) have highly aligned threads, allowing spinons to move freely.
- Insulators have tangled, irregular thread networks, restricting spinon motion.
- Semiconductors have adjustable thread tension properties, enabling controlled conduction.

$$\sigma = \frac{1}{T_{\text{thread}}}$$

where:

- $\sigma$ = Conductivity

- $T_{\text{thread}}$ = Thread resistance to spinon movement

Threads and Spinons Theory
Samir Hanna Safar

- Explains why metals conduct electricity efficiently—thread alignment allows rapid spinon transfer.

- Predicts why superconductors exhibit zero resistance—perfect thread alignment eliminates energy loss.

- Unifies the behavior of conductors, insulators, and semiconductors under one physical model.

4. Resistance – Thread Distortion and Spinon Dissipation

In classical physics, resistance is described as the opposition to electron flow due to collisions with atoms. However, this does not explain:

- Why some materials resist electricity more than others despite having similar atomic structures.

- Why superconductors have zero resistance below critical temperatures.

- How Energy is lost in resistive materials.

Thread- Based Explanation of Resistance:

- Resistance occurs when the thread structure inside a material is disorganized or highly twisted.
- As spinons travel through distorted threads, some of their Energy is converted into heat due to tension misalignment.
- At high temperatures, increased atomic motion disrupts thread alignment, increasing resistance.

$$R = T_{\text{thread}} \cdot L$$

where:

- $R$ = Resistance

- $T_{\text{thread}}$ = Thread distortion factor

- $L$ = Length of the conductive pathway

- Explains why resistance increases with temperature—thermal motion disturbs thread alignment.

- Predicts why resistance depends on material properties—thread structure defines energy dissipation.

- Provides a physical mechanism for superconductivity—perfectly aligned threads allow frictionless spinon movement.

5. The Propagation of Electrical Signals in Circuits

A central paradox in conventional physics is why electric signals travel near the speed of light, even though electrons move slowly.

Thread- Based Explanation of Electrical Propagation:

- When a voltage is applied, spinon energy moves instantly through the thread network, realigning tension across the circuit.
- Electrons remain primarily stationary, oscillating within their localized thread loops, but their spinon interactions relay the signal.

Threads and Spinons Theory
Samir Hanna Safar

- This allows electrical Energy to propagate rapidly, even though individual charge carriers barely move.

- Explains why electrical circuits respond almost instantly—spinon waves travel through the thread network at high speeds.

- Eliminates the need for the "electron drift velocity" paradox—spinons, not electrons, carry the current.

- Unifies electrical propagation with wave mechanics—electricity is a structured thread disturbance, not a particle flow.

6. AC vs. DC – Alternating vs. Continuous Spinon Alignment

Traditional physics describes AC (alternating current) as periodic electron movement and DC (direct current) as continuous unidirectional electron flow. However, this explanation lacks a fundamental physical mechanism.

Thread- Based Explanation of AC and DC:

- In DC, spinons move continuously in one direction through the thread network.
- Spinons oscillate back and forth in AC, creating alternating thread tension waves.
- AC is more efficient for long- distance energy transmission because oscillating tension waves reduce resistance effects.

- Explains why AC voltage is more efficient—thread oscillations minimize spinon dissipation.

- Predicts why high- frequency AC currents exhibit skin effects—spinon motion concentrates in surface threads.

- Unifies AC and DC as variations of spinon alignment in conductive threads.

Conclusion:

A Thread- Based Understanding of Electricity

The Threads and Spinons Theory redefines electrical current as:

- A wave of spinon motion traveling along conductive threads.

- Voltage as a measure of thread tension, driving spinon movement.

- Resistance as thread distortion, disrupting spinon propagation.

- Conductivity as thread alignment, allowing free energy transfer.

- Superconductivity as a state of perfect thread organization, eliminating resistance.

This new framework eliminates outdated particle- based models of electricity and replaces them with a continuous, structured, and mechanically grounded explanation of electrical interactions.

**Chapter 17**

## The Nature of the Magnetic Field in the Threads and Spinons Theory

In traditional physics, the magnetic field is described as a force surrounding moving electric charges and magnetic materials. It is mathematically represented using Maxwell's equations, with field lines indicating the direction and strength of the force. However, this approach lacks a physical explanation of the magnetic field's true nature and how it arises.

Several unanswered questions in conventional physics include:

- What physically constitutes a magnetic field?

- Why do magnetic field lines form loops?

- What causes permanent magnets to retain their magnetism?

- Why does a moving charge generate a magnetic field?

- What is the relationship between magnetism and electricity at a structural level?

The Threads and Spinons Theory provides a new perspective, defining the magnetic field as an accurate, structured alignment of threads and spinons. In this model, magnetism arises from the rotational motion of spinons along thread structures, creating directed energy loops. This explanation eliminates the need for force- carrying particles and provides a mechanically grounded model for magnetism.

This chapter explores the true nature of the magnetic field, how it forms, and how it interacts with charges and currents.

1. The Magnetic Field is a Structured Thread Alignment

Mainstream Assumption:

- Magnetic fields are intangible force fields surrounding moving charges.
- They act at a distance, influencing objects with no direct physical medium.
- Field lines are an abstract way to describe magnetic forces that do not physically exist.

Thread- Based Explanation:

- Magnetic fields are structured thread formations aligning in circular patterns.

- Spinons within threads rotate in synchronized motion, creating a continuous energy loop.
- This alignment generates a directed force, influencing nearby spinons and thread structures.

- Eliminates the need for "force at a distance"—magnetic fields are physical structures.

- Explains why magnetic fields always form loops—spinons align in self- reinforcing patterns.

- Unifies magnetism and electricity as two aspects of thread interactions.

2. How Moving Charges Create a Magnetic Field

A fundamental observation in physics is that moving charges generate a magnetic field. Traditional physics uses Maxwell's equations but does not explain why movement produces magnetism.

Thread- Based Explanation of Charge- Induced Magnetism:

- An electric current is a wave of spinon motion traveling through conductive threads.
- As spinons move, they influence the surrounding thread network, causing nearby threads to align.
- This alignment forms a structured magnetic loop around the current, reinforcing the field.
-

$$B = T_{\text{thread}} \cdot S_{\text{spinon}} \cdot \frac{I}{d}$$

where:

- $B$ = Magnetic field strength
- $T_{\text{thread}}$ = Thread tension
- $S_{\text{spinon}}$ = Spinon rotational energy
- $I$ = Current (spinon flow)
- $d$ = Distance from the current source

- Explains why magnetic fields form around moving charges—spinons reorganize surrounding thread networks.

- Unifies electromagnetism as a natural effect of thread tension and spinon motion.

- Predicts magnetic field strength based on fundamental structural properties of conductive materials.

3. Why Magnetic Field Lines Form Closed Loops

A key feature of magnetism is that magnetic field lines always form closed loops, unlike electric fields, which extend outward. However, traditional physics provides no reason for this.

Thread- Based Explanation of Magnetic Field Loops:

- Spinons in a magnetic field always move in rotational patterns along the threads.
- Since thread networks are continuous, the spinon motion forms a closed- loop structure.

- This self- reinforcing motion maintains a stable, circular magnetic field around currents or magnets.

- Explains why magnetic fields never have a start or end point—spinon motion requires a continuous loop.

- Predicts field strength variations based on thread density and spinon energy.

- Provides an actual physical mechanism for magnetic loops, unlike abstract field line models.

4. The Nature of Permanent Magnets

In classical physics, permanent magnets are explained using quantum spin and electron alignment. However, this model does not explain:

- Why some materials stay magnetized while others do not.

- What physically locks electron spins in place in ferromagnetic materials.

- Why extreme heat or impact can demagnetize materials.

Thread- Based Explanation of Permanent Magnets:

- In magnetic materials, atomic threads are naturally aligned, allowing synchronized spinon motion.
- This alignment creates a self- sustaining magnetic loop, reinforcing the field.
- External forces (heat, impact) disrupt the thread structure, breaking the alignment and demagnetizing the material.

- Explains why only certain materials are magnetic—their threads allow stable spinon alignment.

- Predicts how magnetic domains form—thread structures create self- reinforcing loops.

- Unifies magnetism with an actual mechanical process rather than quantum spin models.

5. Magnetic Induction – How Fields Create Currents

A changing magnetic field can induce an electric current in a nearby conductor, as described by Faraday's Law. However, the exact physical mechanism behind induction remains unclear in traditional physics.

Thread- Based Explanation of Electromagnetic Induction:

- When a magnetic field changes, the spinon alignment in surrounding threads shifts.
- This shift alters the tension within the thread network, transmitting Energy to nearby conductive threads.
- Electrons in the conductor respond by realigning their spinons, generating an electric current.

$$I_{\text{induced}} = \frac{dB}{dt} \cdot T_{\text{thread}} \cdot SEU$$

where:

- $I_{\text{induced}}$ = Induced current
- $dB/dt$ = Rate of change of the magnetic field
- $T_{\text{thread}}$ = Thread network tension
- $SEU$ = Spinon Energy Unit

- Explains induction as a real mechanical effect—thread tension propagates Energy between conductors.

- Unifies electricity and magnetism as interrelated thread structures.

- Provides a deterministic model for induction rather than relying on abstract field descriptions.

6. The Relationship Between Magnetism and Gravity

One of the most mysterious aspects of physics is the possible connection between magnetism and Gravity. Traditional physics treats them as separate forces, but there are unexplained interactions, such as:

Why do strong magnetic fields influence light and particles in ways that resemble gravitational effects?

Why do some cosmic magnetic fields appear to guide matter movement similarly to Gravity?

Threads and Spinons Theory
Samir Hanna Safar

Thread- Based Explanation of Magnetism- Gravity Interaction:

- Both magnetism and Gravity are tension effects within the same universal thread network.
- Gravity arises from large- scale thread contraction, while magnetism is a localized rotational tension effect.
- Magnetic and gravitational effects merge into a unified thread structure in extreme conditions (e.g., neutron stars and black holes).

- Explains why extreme magnetic fields can bend light and particles—both follow thread pathways.

- Predicts potential interactions between magnetism and mass movement, aligning with observed cosmic phenomena.

- Unifies gravity and electromagnetism within a single mechanical framework.

Conclusion:

A Thread- Based Understanding of Magnetism

The Threads and Spinons Theory redefines magnetism as:

A structured alignment of threads and spinons, not an abstract force field. Moving charges induce magnetism by realigning surrounding threads. Permanent magnets retain their fields due to stable spinon loops. Magnetic induction is a tension-driven energy transfer between conductive threads. Gravity and magnetism share a fundamental connection through thread interactions.

Threads and Spinons Theory
Samir Hanna Safar

This new framework eliminates outdated force- based models and replaces them with a mechanically grounded explanation for all magnetic phenomena.

Threads and Spinons Theory
Samir Hanna Safar

**Chapter 18**

**The Nuclear Cocoon**

**A Thread- Based Shield in the Threads and Spinons Theory**

In mainstream physics, atomic nuclei are described as dense collections of protons and neutrons held together by a strong nuclear force. However, this model presents several unresolved questions:

- Why do protons stay bound despite their repulsion?

- Why do neutrons contribute to nuclear stability without having a charge?

- What prevents nuclei from instantly breaking apart under high energy states?

- Why do heavier elements become unstable despite additional neutrons?

The Threads and Spinons Theory introduces the concept of the nuclear cocoon, a structured thread shield that surrounds atomic nuclei. This cocoon:

- Physically protects the nucleus by redistributing thread tension and spinon interactions.

- Explains nuclear stability and decay as a function of thread integrity.

- Eliminates the need for "nuclear force carriers" (gluons), replacing them with thread interactions.

This chapter will define the nuclear cocoon's nature, structure, and function and how it governs atomic stability.

1. What is the Nuclear Cocoon?

The nuclear cocoon is a thread- based energy shield that surrounds and stabilizes an atomic nucleus. It is composed of:

- Interwoven nuclear threads that bind protons and neutrons into a cohesive structure.
- Spinon- aligned tension fields, which distribute nuclear forces evenly.
- A self- reinforcing tension network, preventing external disruption and maintaining stability.

This cocoon is not a physical shell like an electron cloud but a dynamic tension layer that protects and stabilizes the nucleus.

- Explains why nuclear forces act over a short range—cocoon tension is localized.

- Predicts why some nuclei are stable while others decay—cocoon integrity determines longevity.

- Provides a unified structural explanation for nuclear stability and reactions.

2. How the Nuclear Cocoon Prevents Proton Repulsion

Mainstream Assumption:

- The strong nuclear force overcomes proton- proton repulsion, holding the nucleus together.
- This force is more potent than electromagnetism at short distances but does not act beyond nuclear scales.

Thread-Based Explanation:

- The nuclear cocoon prevents proton repulsion by redistributing thread tension evenly.
- Instead of opposing forces acting separately, the cocoon ensures that thread interactions balance out electrostatic repulsion.
- Neutrons stabilize the cocoon by reinforcing thread tension, preventing excessive strain on individual protons.

- Explains why protons stay bound despite repelling charges.

- Predicts why neutron- to- proton ratios influence nuclear stability.

- Eliminates the need for force- carrying gluons—thread mechanics govern stability.

3. The Role of the Nuclear Cocoon in Binding Energy

Conventional physics requires binding energy to separate a nucleus into individual protons and neutrons. However, this model does not explain:

- What physically "stores" this energy within the nucleus?

- Why binding energy varies across different elements.

- How nuclear fusion and fission release energy at a mechanical level.

Thread- Based Explanation of Binding Energy:

- Binding energy is the cumulative tension within the nuclear cocoon.
- Energy is stored in thread compression and spinon realignment inside the cocoon.
- When a nucleus undergoes fusion or fission, cocoon tension is redistributed, releasing stored energy.

$$E_{\text{binding}} = T_{\text{cocoon}} \cdot S_{\text{spinon}} \cdot SEU$$

where:

- $E_{\text{binding}}$ = Nuclear binding energy
- $T_{\text{cocoon}}$ = Tension within the nuclear cocoon
- $S_{\text{spinon}}$ = Spinon alignment factor
- $SEU$ = Spinon Energy Unit (measure of energy transfer capacity)

- Explains why binding energy varies per nucleus—cocoon properties differ per element.

- Predicts energy release mechanisms in nuclear fusion and fission.

- Unifies nuclear energy storage with a mechanical thread-based model.

4. How the Nuclear Cocoon Governs Stability and Decay

In conventional physics, unstable nuclei undergo radioactive decay, emitting alpha, beta, or gamma radiation. However, the reasons behind instability remain unclear.

Thread- Based Explanation of Nuclear Decay:

- A stable nucleus has a well- formed nuclear cocoon that evenly distributes tension.
- If the cocoon becomes strained or misaligned, it begins to break, leading to decay.
- Different types of decay correspond to different thread disruptions:

Alpha decay – A fragment of the cocoon detaches, ejecting a helium nucleus.

Beta decay – A neutron thread collapses, releasing an electron- like wave disturbance.

Gamma decay – Excess spinon energy is emitted as a high- frequency tension wave.

$$\lambda_{\text{decay}} = \frac{T_{\text{cocoon}}}{E_{\text{disruption}}}$$

where:

- $\lambda_{\text{decay}}$ = Probability of decay
- $T_{\text{cocoon}}$ = Nuclear cocoon tension
- $E_{\text{disruption}}$ = Energy required to break thread integrity

- Explains why heavier elements tend to be unstable—larger nuclei stretch the cocoon beyond its natural limits.

- Predicts decay rates based on thread tension rather than probabilistic quantum rules.

- Unifies all types of nuclear decay as structural breakdowns of the cocoon.

5. The Nuclear Cocoon in Fusion and Fission

Fusion – Merging Nuclear Cocoons for Energy Release

- In nuclear fusion, two nuclei must overcome thread tension barriers before merging their cocoons.

- When they fuse, their combined cocoon reorganizes into a lower- energy state, releasing excess thread energy as light or Heat.
- This explains why fusion requires high temperatures—the cocoon must be disrupted to allow merging.

Fission – Breaking the Cocoon to Release Stored Energy

- The cocoon is forcibly split in nuclear fission, causing thread tension to release energy explosively.
- The fission fragments form smaller cocoons with lower tension, stabilizing themselves.
- This model predicts why certain elements (uranium and plutonium) are prone to fission—their cocoons are highly strained.

- Explains why fusion and fission release energy—the cocoon redistributes tension into lower- energy configurations.

- Predicts energy yield based on cocoon properties rather than quantum probability.

- Unifies all nuclear reactions under a single mechanical thread- based framework.

6. Applications of the Nuclear Cocoon Concept

The nuclear cocoon model has profound implications for:

- Nuclear energy research – Designing materials that control cocoon integrity for safer reactors.

- Particle physics – Understanding subatomic structure through thread interactions rather than quarks and gluons.
- Astrophysics – Explaining stellar nucleosynthesis as cocoon- driven fusion.
- Medical applications – Developing targeted radiation treatments by manipulating cocoon properties of unstable isotopes.

- Offers real- world applications in multiple scientific fields.

- Provides a mechanical basis for nuclear behavior beyond the Standard Model.

- Eliminates reliance on force- carrying particles, unifying nuclear physics under a tangible thread structure.

Conclusion:

A New Understanding of Nuclear Structure

The Threads and Spinons Theory redefines the atomic nucleus as:

- Surrounded by a structured nuclear cocoon that governs stability.

- Protected from external forces by thread tension and spinon alignment.

- Held together without force- carrying particles—thread interactions control nuclear integrity.

- Undergoing decay when cocoon integrity is disrupted.

- Releasing energy in fusion and fission through cocoon tension redistribution.

This revolutionary model replaces the Standard Model's force-based approach with an actual mechanical structure, explaining all nuclear interactions with deterministic thread mechanics.

Chapter 19

## The Outer Atomic Surface – A Thread- Based Model of Atomic Boundaries in the Threads and Spinons Theory

In mainstream atomic Theory, the outer surface of an atom is described as an electron cloud, where electrons exist in probabilistic orbitals around the nucleus. This model, based on quantum mechanics, introduces several conceptual and physical challenges:

- Why do electrons remain confined around the nucleus instead of escaping?

- What physically defines the outer boundary of an atom?

- Why do atoms have defined sizes despite the probabilistic nature of electron positions?

- How do atoms maintain structural integrity when interacting with other atoms?

The Threads and Spinons Theory introduces a new concept: the outer atomic surface is not an electron cloud but a structured layer of thread tension that forms a dynamic, protective boundary around the atom.

This chapter will define the outer atomic surface in extensive detail and explore its nature, formation, function, and role in atomic interactions.

1. What is the Outer Atomic Surface?

The outer atomic surface is a tensioned thread shell that encapsulates the nucleus and electrons within a structured energy field. It consists of:

- Interwoven atomic threads that form a structured boundary.

- Spinon- driven dynamic motion, reinforcing the shell's stability.

- A tension- stabilized interface that prevents atoms from collapsing or dispersing.

This surface is not an electron cloud but a mechanically stable, thread- based protective layer that defines an atom's physical extent.

- Explains why atoms have defined sizes—thread tension determines atomic boundaries.

Threads and Spinons Theory
Samir Hanna Safar

- Unifies atomic structure with an actual mechanical model, eliminating wavefunction uncertainties.

- Predicts atomic stability based on thread integrity rather than quantum probabilities.

2. Why Electrons Do Not Escape – The Outer Shell as a Confinement Mechanism

Mainstream Assumption:

- Electrons are bound to the nucleus by electrostatic attraction but move freely in probability clouds.
- The Heisenberg uncertainty principle prevents electrons from having fixed positions.

Thread- Based Explanation:

- Electrons are not free- floating particles but structured thread loops embedded within the atomic surface.
- The outer atomic shell acts as a tension barrier, preventing electrons from escaping.
- Spinon alignment within the shell maintains electron stability, ensuring confinement.

- Explains why electrons remain in stable positions—thread-based confinement replaces abstract wavefunctions.

- Predicts energy levels as actual mechanical structures, not probability zones.

- Unifies atomic integrity with a physically defined surface, eliminating quantum paradoxes.

3. Atomic Size and the Role of Thread Tension

In traditional atomic Theory, atomic size is determined by electron cloud distribution, which lacks a physical mechanism for defining rigid boundaries.

Thread- Based Explanation of Atomic Size:

- The radius of an atom is determined by the maximum thread extension that maintains stable tension.
- Firmer nuclear tension pulls the outer shell inward, resulting in smaller atomic sizes.
- Weaker nuclear tension allows the outer shell to expand, increasing atomic size.

$$R_{atom} = \frac{T_{outer\ shell}}{T_{nucleus}} \cdot SEU$$

where:

- $R_{atom}$ = Atomic radius
- $T_{outer\ shell}$ = Thread tension at the atomic boundary
- $T_{nucleus}$ = Nuclear thread tension
- $SEU$ = Spinon Energy Unit

- Explains why atomic size follows periodic trends—tension levels dictate expansion or contraction.

- Predicts atomic radii based on thread structure rather than probabilistic charge distributions.

- Unifies atomic size determination with real mechanical forces.

Threads and Spinons Theory
Samir Hanna Safar

4. The Outer Atomic Surface as a Protective Barrier

Atoms do not collapse or merge uncontrollably, even nearby, suggesting a protective mechanism at their outer boundaries. Traditional physics attributes this to electron repulsion, but no clear explanation is given for why atomic surfaces resist compression.

Thread- Based Explanation of Atomic Surface Protection:

- The outer shell forms a dynamic tension barrier that resists direct compression.
- Spinons in the shell create a repelling force when two atoms approach, preventing collapse.
- Attractive forces only engage when thread structures align favorably, allowing bonding.

- Explains why atoms do not collapse into each other—thread tension prevents direct merging.

- Predicts interatomic distances in materials based on shell properties.

- Eliminates the need for repulsive quantum exclusion principles—mechanical tension governs atomic integrity.

5. How Atoms Bond Without Overlapping Outer Surfaces

In classical chemistry, atomic bonds are explained using electron sharing (covalent), electron transfer (ionic), or metallic delocalization. However, these explanations do not address why:

- Atoms never physically merge into one another.

Threads and Spinons Theory
Samir Hanna Safar

- Electron orbitals do not collapse into a single structure upon bonding.

- Bonds maintain fixed lengths instead of being fluid or highly variable.

Thread- Based Explanation of Atomic Bonding:

- Atoms never fully merge because their outer thread surfaces maintain structural integrity.
- Bonding occurs when a thread loops between two atoms interlock, forming a stable energy bridge.
- The strength and length of the bond depend on how well the thread loops align between atoms.

- Explains why bond lengths are fixed—thread configurations determine stable distances.

- Predicts why different elements form specific bonds—thread compatibility governs bonding potential.

- Unifies all chemical interactions under a single mechanical thread structure.

6. The Outer Atomic Surface and Electromagnetic Interactions

Atoms interact with light, electrical, and magnetic fields, but conventional physics does not explain why some interactions penetrate the atom while others only affect the surface.

Thread- Based Explanation of Electromagnetic Interactions:

- The outer atomic shell acts as a conductive surface, allowing spinon- induced interactions with electromagnetic waves.
- Light absorption occurs when external spinon waves match the shell's natural vibrational frequency.
- Electrical Conduction occurs when spinons transfer energy through the shell, affecting internal thread structures.

- Explains why light absorption is element- dependent—thread properties define energy resonance.

- Predicts why metallic atoms conduct electricity—thread structures allow free spinon movement.

- Unifies electromagnetic interactions with accurate mechanical thread alignments.

7. The Outer Atomic Surface and Atomic Excitation

In classical physics, atomic excitation is described as electrons absorbing energy and moving to higher orbitals. However, this does not explain:

- How atoms physically store and release energy.

- Why excited atoms return to lower energy states.

- Why energy levels are quantized rather than continuous.

Thread- Based Explanation of Atomic Excitation:

- Excited atoms store energy as increased tension in the outer shell threads.

- Spinons temporarily alter their alignment, increasing vibrational energy.
- When the atom relaxes, spinons return to a lower energy state, releasing energy as electromagnetic Radiation.

- Explains why atomic energy levels are discrete—thread configurations allow only specific resonances.

- Predicts spectral lines based on thread oscillation frequencies.

- Eliminates quantum uncertainty—excitation is a structured thread- based process.

Conclusion:

A Thread- Based Model of the Outer Atomic Surface

The Threads and Spinons Theory redefines the outer atomic surface as:

- A structured, tension- stabilized thread network enclosing the atom.

- A protective barrier that prevents electron escape and atomic collapse.

- The primary interface for atomic bonding and electromagnetic interactions.

- A mechanism for defining atomic size, structure, and excitation properties.

Threads and Spinons Theory
Samir Hanna Safar

- A unifying concept replacing quantum probability clouds with real mechanical constraints.

This new perspective replaces the uncertain, wavefunction-based atomic model with a deterministic, structured framework for understanding atomic integrity.

The next chapter will explore how gravity emerges from the thread network, replacing space- time curvature with fundamental mechanical interactions.

Threads and Spinons Theory
Samir Hanna Safar

Threads and Spinons Theory
Samir Hanna Safar

**Chapter 20**

**Energy Transfer and Heat Transfer in the Threads and
Spinons Theory**

In classical physics, energy transfer and heat transfer
are described as the movement of kinetic energy from
one particle to another through Conduction, Convection,
or Radiation. However, these explanations raise several
unanswered questions:

- What physically happens when energy moves from one
object to another?

- Why do molecules absorb and release Heat in quantized
steps?

Threads and Spinons Theory
Samir Hanna Safar

- What fundamentally distinguishes different types of heat transfer?

- How does Heat interact with materials at the atomic level?

The Threads and Spinons Theory provides a mechanical explanation for heat transfer by defining energy as structured spinon motion within the universal thread network. This Theory eliminates the random motion- based view of energy transfer and replaces it with a continuous tension and spinon alignment process that dictates how Heat moves through materials.

This chapter will explore the true nature of energy transfer and heat transfer, using the example of boiling water to illustrate the Theory in action.

1. What is Energy Transfer in the Threads and Spinons Theory?

Mainstream Assumption:

- Energy is the ability to work, moving as kinetic energy from one particle to another.
- Heat transfer occurs through particle collisions, wave emissions, or molecular vibrations.

Thread- Based Explanation:

- Energy is not just kinetic motion but the structured movement of spinons through the thread network.
- Energy transfer occurs when spinon waves propagate along threads, altering their tension and motion.

- Heat is a specific form of energy transfer where spinons realign within atomic and molecular threads.

- Explains why energy moves in discrete amounts—spinon stacks transfer in quantized steps.

- Predicts energy transfer efficiency based on thread structure and spinon alignment.

- Unifies all forms of energy transfer under a single physical model.

2. The Three Types of Heat Transfer in Thread- Based Mechanics

1. Conduction – Direct Thread Tension Propagation

Example: Heating a metal rod from one end

In classical physics, Conduction is explained as the transfer of kinetic energy from fast- moving molecules to slower-moving ones through collisions.

Thread-Based Explanation of Conduction:

- Conduction occurs when thread tension shifts through direct atomic connections.
- Spinons transfer energy along the thread network, realigning molecular threads.
- This propagates a structured wave of spinon motion, carrying Heat through the material.

$$Q_{\text{conduction}} = k \cdot T_{\text{thread}} \cdot SEU$$

where:

- $Q_{\text{conduction}}$ = Heat transferred by conduction
- $k$ = Thermal conductivity (determined by thread alignment)
- $T_{\text{thread}}$ = Thread tension gradient
- $SEU$ = Spinon Energy Unit

- Explains why metals conduct Heat efficiently—highly aligned threads allow rapid spinon propagation.

- Predicts material- dependent conductivity based on thread structure, not atomic spacing.

- Unifies conduction as a structured spinon energy transfer process.

2. Convection – Spinon Wave Propagation in Fluids

Example: Boiling water in a pot

In classical physics, Convection is explained as the movement of Heat through bulk fluid motion, where hotter, less dense regions rise while cooler, denser regions sink.

Thread- Based Explanation of Convection:

- Convection occurs when spinons generate thread tension gradients in a fluid, influencing atomic motion.
- Thread realignment alters density distributions, causing fluid movement.

Threads and Spinons Theory
Samir Hanna Safar

- Spinon wave propagation dictates heat movement patterns, forming convection currents.

$$Q_{\text{convection}} = h \cdot \Delta T_{\text{fluid}} \cdot SEU$$

where:

- $Q_{\text{convection}}$ = Heat transferred by convection
- $h$ = Heat transfer coefficient (related to spinon wave efficiency)
- $\Delta T_{\text{fluid}}$ = Temperature difference driving convection
- $SEU$ = Spinon Energy Unit

- Explains why convection currents form naturally—spinon realignment controls density variations.

- Predicts how Convection occurs in gases, liquids, and plasmas based on spinon behavior.

- Unifies convection as a spinon wave- driven process, not just bulk motion.

3. Radiation – Spinon Wave Emission Through Free Threads

Example: The Heat felt from boiling water without touching it

In classical physics, Radiation is described as electromagnetic waves carrying Heat away from an object. However, this explanation does not define:

- What physically "emits" this energy from the surface?

Threads and Spinons Theory
Samir Hanna Safar

- How do Radiation does on moves move space without a medium?

- Why does Radiation interact differently with different materials?

Thread- Based Explanation of Radiation:

- Radiation is the release of spinon waves into the surrounding free- thread network.
- As a hot object's spinons realign, excess energy is emitted outward as structured waves.
- These waves propagate through the universal thread medium, transferring energy to other objects:

$$Q_{\text{radiation}} = \sigma \cdot T^4_{\text{surface}} \cdot SEU$$

where:

- $Q_{\text{radiation}}$ = Heat transferred by radiation
- $\sigma$ = Stefan-Boltzmann constant (related to spinon emission efficiency)
- $T_{\text{surface}}$ = Temperature of the emitting surface
- $SEU$ = Spinon Energy Unit

- Explains why Radiation moves through space—threads extend everywhere, forming a medium.

- Predicts absorption and reflection properties based on material thread alignment.

Threads and Spinons Theory
Samir Hanna Safar

- Unifies radiation as a structured spinon emission, not wave-particle duality.

3. Example: Boiling Water in a Pot

Let us apply the Threads and Spinons Theory to boiling water to explain heat transfer in real- world conditions.

1 . Conduction: Heat from the stove transfers through the metal pot via thread tension waves in the material. The spinons in the metal threads align, transferring energy to the water.

2 . Convection: As water molecules absorb spinon waves, thread realignment reduces density in heated regions, creating convection currents that circulate Heat.

3 . Radiation: The boiling water's outer atomic shell releases excess spinon energy, emitting Heat as infrared Radiation that can be felt without touching the pot.

- Explains all aspects of boiling water without relying on separate force- based models.

- Predicts how energy transfer changes based on material properties.

- Unifies conduction, Convection, and Radiation under one thread- based framework.

4. The Relationship Between Heat and Temperature

In conventional physics, Temperature is defined as the average kinetic energy of particles, but this explanation does not clarify why:

- Heat and Temperature are not always proportional (e.g., phase changes).

- Different materials store and release Heat at different rates.

- Extreme Heat does not always increase particle motion (e.g., plasma states).

Thread- Based Explanation of Temperature:

- Temperature is the density of spinon motion within a material's thread network.
- Heat is the process of redistributing spinon alignment to increase or decrease Temperature.
- Phase changes occur when thread tension forces atomic restructuring, absorbing or releasing latent energy.

- Explains why temperature plateaus during phase changes— thread realignment absorbs energy.

- Predicts material- specific heat capacities based on thread structure.

- Unifies heat and Temperature as spinon density effects, not just kinetic energy.

Conclusion:

A Thread- Based Understanding of Heat Transfer

Threads and Spinons Theory
Samir Hanna Safar

The Threads and Spinons Theory redefines Heat and energy transfer as:

- Spinon- driven tension waves propagating through atomic threads.

- Conduction as direct thread tension redistribution.

- Convection as spinon wave- induced fluid motion.

- Radiation as free- thread spinon emission.

- Temperature as a function of spinon density and motion.

This new framework replaces random motion- based thermodynamics with structured mechanical interactions, providing a clear and unified explanation of energy transfer.

Threads and Spinons Theory
Samir Hanna Safar

# Part III

# Experiments and Quantum Phenomena

Threads and Spinons Theory
Samir Hanna Safar

## Chapter 21

### The Double- Slit Experiment

### A Thread- Based Explanation Without Paradoxes

The double- slit experiment is one of the most famous and perplexing experiments in quantum mechanics. It demonstrates that light and particles behave as both waves and particles, creating an interference pattern even when sent through the slits one at a time. The experiment has led to wave- particle duality, the Copenhagen interpretation, and even ideas about observer- dependent reality in quantum mechanics.

However, these explanations raise fundamental paradoxes and unresolved questions:

- Why does a single particle produce an interference pattern as if it travels through both slits?

- Why does the interference pattern disappear when we observe which slit the particle passes through?

- What physically causes the wave- like behavior of particles in the first place?

- Why does a detector affect the outcome when it does not physically block the path?

The Threads and Spinons Theory eliminates these paradoxes. The interference pattern is not a result of quantum probability or duality but is caused by fundamental physical structures— threads and spinons—that guide particles along tensioned pathways.

This chapter will retrieve and expand on our previous analysis of the double- slit experiment, providing a clear, deterministic, and paradox- free explanation for the observed phenomena.

1. The Double- Slit Experiment: Setup and Observations

The Classical Experiment

A light source or particle emitter sends photons, electrons, or even atoms toward a barrier with two narrow slits. On the other side, a detection screen records the impact pattern.

Key Observations:

1. When both slits are open, an interference pattern appears, even if particles are sent one at a time.

Threads and Spinons Theory
Samir Hanna Safar

2. When one slit is blocked, no interference pattern appears—
only a single- slit diffraction pattern.

3□. When detectors are placed to observe which slit a particle
goes through, the interference pattern disappears.

4□. The interference pattern remains Even if we use low-
energy particles (like large molecules).

Conventional Explanation:

- Particles behave as probabilistic waves, traveling
  through both slits and interfering with themselves.
- Observation "collapses the wavefunction", forcing a
  single outcome.
- Quantum mechanics claims reality is indeterminate
  until measured.

- Problems with this explanation:

- Why does an individual particle create an interference
  pattern?
- Why does Observation change the outcome?
- How does a single particle "interfere with itself"
  without splitting?

2. The Threads and Spinons Theory Explanation

In this Theory, electrons and photons are not point particles in
probability waves. Instead, they are actual physical thread
structures with spinons traveling along them.

- An electron is a loosely bound cloud of thread loops and
spinons.

- It follows a deterministic path guided by the underlying thread network.

- The interference pattern results from thread realignment and spinon synchronization, not probability waves.

3. How the Interference Pattern is Formed

Step 1: Particle Movement is Guided by Threads

- When a photon or electron is emitted, it follows a pre-existing network of stretched threads, forming a path between the emitter and the screen.
- These threads are already tensioned and interacting with the environment.
- The particle's spinons align with these threads, dictating its motion.

Step 2: Threads Passing Through Both Slits Influence the Particle's Path

- As the particle approaches the slits, the pre- existing threads split into two main pathways.
- Even though the particle itself only goes through one slit, its thread connections extend through both slits simultaneously.
- The spinon motion within the threads interacts with itself on both paths, causing realignment before reaching the screen.

Step 3: Wave- Like Motion Emerges Naturally from Thread Interactions

- The spinons in the threads oscillate, producing a structured interference pattern.
- When the particle reaches the screen, it lands where the thread tension guides it, not randomly.
- The resulting pattern is not a "wave function collapse" but a structured pathway alignment created by the thread network.

- No wave- particle duality needed—thread tension naturally produces the pattern.

- No need for probabilistic interpretations—particles follow deterministic pathways.

- No particle "splitting"—only thread- based interactions affect the outcome.

4. Why the Pattern Disappears When Observed

A key mystery is why the interference pattern disappears when we measure which slit the particle goes through. According to quantum mechanics, measurement "collapses the wavefunction."

Thread- Based Explanation:

- When a detector is placed, it physically interacts with the pre- existing threads.
- This disturbs the spinon alignment, breaking the delicate interference condition.
- Instead of an organized pattern, the particle follows a disrupted path, producing two bands on the screen.

- Observation does not collapse probability—it mechanically disturbs the thread network.

- Predicts that any physical disturbance (not just measurement) will break the pattern.

- Unifies the observer effect with real physical interaction—no need for metaphysical explanations.

5. Why Even Large Particles Show Interference

Experiments have shown that even molecules as large as buckyballs (C60) produce interference patterns, challenging conventional quantum mechanics.

Thread- Based Explanation:

- Larger molecules still have thread connections extending through both slits.
- Their internal spinons still interact with the thread network, forming an interference pattern.
- The pattern disappears if the molecule is large enough to disrupt the network due to excessive thread tension.

- Explains why interference persists for large particles—their threads remain connected through both slits.

- Predicts an upper limit where interference stops—once the molecule disrupts the network too much, no pattern forms.

- Unifies all matter wave behavior as a function of thread structure and tension.

## 5. How This Theory Resolves All Double- Slit Paradoxes

| Quantum Mechanics Issue | Thread-Based Explanation |
| --- | --- |
| How does a particle interfere with itself? | It doesn't—its threads interact through both slits, guiding its motion. |
| Why does the interference pattern disappear when observed? | Measurement physically disturbs the thread network, breaking interference. |
| Why does a single particle follow a probabilistic path? | It doesn't—its path is determined by pre-existing thread structures. |
| Why does interference occur even when particles are sent one at a time? | Each particle connects to both slits through its thread network, maintaining interference. |
| Why do even large molecules show interference? | Their internal thread structures still interact with the slits, creating wave-like motion. |

- No probability waves, no observer- dependent reality—only fundamental physical interactions.

- Interference patterns arise from structured spinon pathways, not wave function superposition.

- Measurement effects are due to mechanical disruption, not metaphysical wave collapse.

Conclusion:

The Double- Slit Experiment as a Real Physical Effect

The Threads and Spinons Theory redefines the double- slit experiment as:

A deterministic process where particles follow structured thread pathways. An interference pattern created by fundamental spinon interactions in the thread network. An observer effect caused by thread disruption, not probability

Threads and Spinons Theory
Samir Hanna Safar

collapse. A natural consequence of thread tension guiding all particle motion.

This new perspective eliminates all quantum paradoxes, providing a mechanically grounded explanation for the double- slit experiment without requiring wave- particle duality, probability wavefunctions, or observer- based reality.

# Chapter 22

## Testing the Double- Slit Experiment with Electrons and Protons

In the Threads and Spinons Theory, the double- slit experiment is explained by actual thread structures guiding particle motion, rather than wave- particle duality or probabilistic behavior. We have already shown how the interference pattern is formed by spinon motion and thread tension interactions. Now, we will expand on this by analyzing what happens when we perform the double- slit experiment using different particles—electrons and protons.

- Electrons and protons are not point particles but thread- based structures with spinon motion.

- The behavior of these particles in the double- slit experiment is determined by how their internal threads interact with the slit edges and the pre- existing thread network.

Threads and Spinons Theory
Samir Hanna Safar

- Each type of particle interacts with the double slit in a unique way due to its structural properties.

## 1. Shooting an Electron Through the Double- Slit Experiment

Electrons have been widely used in the double- slit experiment, producing interference patterns similar to photons. However, in conventional quantum mechanics, this is attributed to wave- particle duality and wavefunction collapse, which contradicts classical intuition.

Thread- Based Explanation of Electron Interference

- An electron is a loose cloud of thread loops and spinons.

- When an electron is emitted, it follows the pre- existing thread network, forming a path to the detection screen.

- Its thread loops extend outward, interacting with both slits at the same time, even if the electron itself passes through only one slit.

- Spinon motion within the thread structure aligns with both slit openings, creating the interference effect.

Threads and Spinons Theory
Samir Hanna Safar

$$I_{\text{electron}} = \frac{S_{\text{spinon}}}{T_{\text{thread}}} \cdot SEU$$

where:

- $I_{\text{electron}}$ = Electron interference pattern strength
- $S_{\text{spinon}}$ = Spinon motion in the electron's thread structure
- $T_{\text{thread}}$ = Thread network tension near the slits
- $SEU$ = Spinon Energy Unit

Key Predictions for Electrons in the Double- Slit Experiment

- Electrons will always produce an interference pattern, even when fired one at a time because their thread loops interact with both slits.

- If an external detector disturbs the thread network, the interference pattern will collapse, producing only two bands.

- Electrons with higher energy (faster movement) will produce slightly tighter interference bands due to increased thread tension.

Experiment:

- Firing electrons at different speeds should produce different interference bandwidths due to changes in thread tension.

2. Shooting a Proton Through the Double- Slit Experiment

What Happens When We Fire a Proton?

Protons are much more massive than electrons and have a tighter internal thread structure. Unlike electrons, which are loosely bound clouds of threads, protons have denser spinon configurations, making their interactions with thread networks different.

- A proton is a tightly coiled thread structure held together by strong internal tension.

- Its spinon motion is more rigid, meaning it interacts differently with the slits.

- Instead of forming a widespread interference pattern like an electron, its interaction with slits is more constrained.

Thread- Based Explanation of Proton Behavior

- Protons still have thread loops extending beyond their primary structure, but these loops are much smaller than those of electrons.

- As a result, a proton's threads interact less with both slits simultaneously, reducing interference effects.

- The interference pattern should still form but with lower contrast and broader spacing than in the electron case.

$$I_{\text{proton}} = \frac{S_{\text{spinon}}}{T_{\text{thread}}} \cdot SEU \cdot \frac{1}{m_p}$$

where:

- $I_{\text{proton}}$ = Proton interference pattern strength

- $S_{\text{spinon}}$ = Spinon motion in the proton's thread structure

- $T_{\text{thread}}$ = Thread network tension near the slits

- $SEU$ = Spinon Energy Unit

- $m_p$ = Proton mass (higher mass reduces interference effects)

## Key Predictions for Protons in the Double- Slit Experiment

- Protons should still exhibit an interference pattern, but they will be less pronounced than electrons.

- The distance between interference bands will be wider due to the proton's higher mass and tighter thread configuration.

- Protons moving at higher speeds should show less interference as their spinon alignment stiffens.

Experiment:

- Compare interference patterns for protons at different velocities—faster protons should show weaker interference.

### 3. Comparison of Electron and Proton Result

| Particle | Thread Structure | Spinon Motion | Expected Interference | Effect of Speed Increase |
|---|---|---|---|---|
| Electron | Loose thread cloud | High flexibility | Strong interference pattern | Tighter bands, stronger effect |
| Proton | Tightly coiled structure | Lower flexibility | Weak but present interference | Bands spread apart, weaker effect |

Threads and Spinons Theory
Samir Hanna Safar

General Findings

- Electrons show more substantial interference because of their flexible thread structure.

- Protons show weaker interference due to their dense, tightly wound threads.

- The interference effect weakens as mass increases and thread flexibility decreases.

This suggests that the interference effect is not caused by quantum probability but by how a particle's threads interact with the slit structure.

4. What Happens If We Fire Larger Particles?

Now, let us consider what happens when we use even larger particles, like heavy atoms or molecules (e.g., buckyballs).

- The more significant the particle, the denser and more rigid its internal thread structure becomes.

- At a specific size, the threads become too short to extend through both slits simultaneously.

- This means interference should disappear completely for massive particles.

- If a particle's thread structure is too rigid to interact with both slits, it will behave like a classical object and not show interference.

- There exists a threshold where interference stops because the particle's thread connections no longer extend far enough.

Experiment:

- Fire molecules of increasing size through the double slit and observe when the interference pattern vanishes.

### 6. How This Theory Resolves All Double-Slit Paradoxes for Different Particles

| Quantum Mechanics Issue | Thread-Based Explanation |
| --- | --- |
| Why does an electron interfere with itself? | It doesn't—its threads interact through both slits, guiding its motion. |
| Why does the interference pattern disappear when observed? | Measurement physically disturbs the thread network, breaking interference. |
| Why does a single particle follow a probabilistic path? | It doesn't—its path is determined by pre-existing thread structures. |
| Why do larger particles show less interference? | Their threads are too rigid to extend through both slits, reducing interference. |
| Why does a proton behave differently than an electron? | Its thread structure is tighter, limiting interference effects. |
| | ↓ |

- No wave- particle duality needed—thread tension naturally produces the pattern.

- No need for probabilistic interpretations—particles follow deterministic pathways.

- No particle "splitting"—only thread- based interactions affect the outcome.

Conclusion: The Double- Slit Experiment as a Real Physical Effect

Threads and Spinons Theory
Samir Hanna Safar

The Threads and Spinons Theory redefines the double- slit experiment as:

- A deterministic process where particles follow structured thread pathways.

- An interference pattern created by fundamental spinon interactions in the thread network.

- An observer effect caused by thread disruption, not probability collapse.

- A natural consequence of thread tension guiding all particle motion.

This new perspective eliminates all quantum paradoxes, providing a mechanically grounded explanation for the double-slit experiment without requiring wave- particle duality, probability wavefunctions, or observer-based reality.

Chapter 23

## The Schrödinger's Cat Paradox

## A Thread- Based Rebuttal to the Copenhagen Interpretation

The Schrödinger's cat paradox is one of the most famous and controversial thought experiments in quantum mechanics. It was initially proposed by Erwin Schrödinger in 1935 to highlight the absurdity of the Copenhagen interpretation, which suggests that quantum systems exist in multiple states (superpositions) until observed.

The Thought Experiment:

1□. A cat is placed inside a sealed box.

2□. Inside the box is a radioactive atom with a 50% chance of decaying within a given timeframe.

3□. If the atom decays, it triggers a poison release, killing the cat.

4□. If the atom does not decay, the cat remains alive.

5□. According to Copenhagen's interpretation, the cat is dead and alive until the box is opened; at this point, the wavefunction "collapses" into a definite state.

1. The Copenhagen Interpretation and Its Flaws

The Copenhagen interpretation, developed by Niels Bohr and Werner Heisenberg, is one of the oldest and most widely taught interpretations of quantum mechanics. It states that:

- Quantum systems exist in a superposition of multiple states until observed.

- Observation collapses the wavefunction, forcing a definite outcome.

- Reality is inherently probabilistic, with no deterministic behavior at small scales.

Flaws in the Copenhagen Interpretation:

- It suggests that Reality does not exist until observed.

- It treats Observation as a mystical process that "forces" Reality into one outcome.

Threads and Spinons Theory
Samir Hanna Safar

- It fails to explain why macroscopic objects (like a cat) should follow quantum Superposition.

- It introduces paradoxes like Wigner's Friend (where one observer sees a different reality than another).

The Threads and Spinons Theory rejects the Copenhagen interpretation entirely and provides a mechanical, deterministic explanation of Schrödinger's cat paradox.

2. A Thread- Based Explanation of the "Dead and Alive" Paradox

Key Assumptions in Threads and Spinons Theory:

- All objects, including atoms, cats, and humans, are connected through physical thread structures.

- Spinons travel along these threads, transmitting information and energy deterministically.

- There is no "wavefunction collapse"—only thread interactions that determine outcomes.

- Observation does not create Reality; it simply detects pre-existing conditions.

Why the Cat is Never in a Superposition

- The radioactive atom follows a deterministic spinon energy transfer mechanism, not quantum randomness.

- If the atom decays, it happens due to spinon motion disrupting thread tension—not probability waves.

Threads and Spinons Theory
Samir Hanna Safar

- The cat's state (alive or dead) is determined when the spinon interaction occurs.

- The observer does not affect Reality—the outcome is fixed based on the atom's accurate energy transfer.

- There is no dual- state cat—just one fixed Reality.

- The "collapse" of the wavefunction is an illusion—threads already determined the outcome.

- Observation does not change Reality; it only reveals what threads have already structured.

Mathematical Representation of Deterministic Decay

$$P_{\text{decay}} = \frac{T_{\text{thread}}}{S_{\text{spinon}}} \cdot SEU$$

where:

- $P_{\text{decay}}$ = Probability of decay (but deterministically calculated)
- $T_{\text{thread}}$ = Thread tension in the nucleus
- $S_{\text{spinon}}$ = Spinon alignment probability
- $SEU$ = Spinon Energy Unit

- If the atom decays, the poison is released. If it does not decay, the cat remains alive.

- The cat's state is already set, whether observed or not.

- There is no superposition, only a deterministic chain of spinon- guided events.

Threads and Spinons Theory
Samir Hanna Safar

- Observation is irrelevant—thread tension and spinon activity determine the cat's state.

- No reality splitting or parallel universes—only structured mechanical interactions.

## 3. How This Theory Negates the Copenhagen Interpretation

| Copenhagen Interpretation (Quantum Mechanics) | Threads and Spinons Theory (Deterministic Physics) |
| --- | --- |
| The cat is in a **superposition of both alive and dead** until measured. | The cat is **either alive or dead from the moment of decay** —no superposition exists. |
| Wavefunction collapse happens **upon observation.** | There is **no wavefunction collapse**—only deterministic thread interactions. |
| The outcome is based on **random probability.** | The outcome is **determined by spinon interactions along threads.** |
| Reality does not exist until it is observed. | Reality is **pre-existing and does not depend on observation.** |
| Particles exist in probabilistic states. | Particles follow **structured pathways based on thread tension and spinon alignment.** |

- No paradox—Schrödinger's cat has one definite fate, whether observed or not.

- Observation is passive—it does not force Reality to choose an outcome.

- Macroscopic objects do not follow quantum Superposition—this was a flawed assumption.

## 4. Addressing Other Quantum Paradoxes with Threads and Spinons Theory

### A. Wigner's Friend Paradox

- Traditional View:

- One observer inside a lab sees a definite outcome (e.g., a decayed atom).
- Another observer outside the lab believes the system is still in Superposition.
- This creates a paradox where two people see different realities.

- Thread- Based Explanation:

- The decay event is already determined by spinon movement.
- The second observer outside the lab is unaware of the event, but the event has already occurred.
- There is no multiple- reality paradox—only one fixed sequence of events.

B. The Quantum Measurement Problem

- Traditional View:

- Measuring a system collapses its wavefunction.
- Before Measurement, Reality is undefined.

- Thread- Based Explanation:

- Measurement does not collapse anything—it only detects the existing state.
- The particle's motion was set by thread tension and spinon behavior long before the Measurement.
- There is no need for probabilistic interpretations.

5. Implications of Rejecting Copenhagen Interpretation

- Quantum mechanics still works—but it is incomplete.

- Thread- based interactions govern all atomic behavior deterministically.

- Superposition is an illusion created by misunderstanding spinon interactions.

- There is no observer- dependent reality—Reality exists independent of Measurement.

\* Future Experiment to Confirm the Theory

- Design a real- time tracking system to observe decay before and after "wavefunction collapse."

- If decay is already set before Measurement, it confirms the thread- based deterministic model.

- Compare the results with quantum superposition models and predict deviations based on thread interactions.

Conclusion:

Reality is Deterministic, Not Probabilistic

The Threads and Spinons Theory redefines Schrödinger's cat and the Copenhagen interpretation as:

A deterministic process where thread tension and spinon motion dictate atomic events. A rejection of quantum randomness in favor of structured physics. An observer-independent reality, where Measurement only reveals, not creates, an outcome. A framework that resolves all paradoxes, eliminating the need for wavefunction collapse. Quantum physics does not need mysterious probability waves or superpositions—it needs a real physical foundation, and the Threads and Spinons Theory provides it.

**Chapter 23-B**

## Quantum Entanglement

### A Thread- Based Explanation for "Spooky Action at a Distance"

Quantum Entanglement is one of modern physics strangest and most controversial phenomena. Albert Einstein famously described it as "spooky action at a distance" because two entangled particles appear to influence each other instantaneously, no matter how far apart they are.T

The Entanglement Experiment

In a typical quantum entanglement experiment:

A pair of particles (e.g., photons or electrons) is created in a controlled system.

Vast distances, then separate these particles.

Threads and Spinons Theory
Samir Hanna Safar

When one particle is measured, its partner instantly "knows" the measurement outcome and adjusts accordingly.

Example: If two entangled electrons are sent in opposite directions and we measure one's spin as "up," the other's spin will always be "down"—even if they are light- years apart!

Problems with the Mainstream Explanation of Entanglement:

In standard quantum mechanics, Entanglement is described using the Copenhagen Interpretation and the mathematical formalism of wavefunctions:

- Entangled particles exist in a superposition until measured.

- When one particle is measured, its wavefunction collapses, and the other instantly takes on the opposite state.

- This happens faster than the speed of Light, violating classical physics!

Flaws in this Explanation:

- It does not explain how information is transmitted instantaneously.

- It suggests "non- locality" without a physical mechanism.

- It treats Measurement as a mystical process that "collapses" Reality.

- It does not define what physically connects entangled particles across space.

The Threads and Spinons Theory provides a more straightforward, mechanical explanation for quantum Entanglement—without paradoxes.

The Thread- Based Explanation of Quantum Entanglement:

- Entangled particles remain connected by actual physical threads, even when separated.

- These threads stretch across space but remain under tension.

- Spinons travel along these threads, synchronizing the particles' states in real Time.

- There is no "instantaneous collapse"—just a mechanical transfer of information.

- How Entanglement Works:

- When a pair of particles is created, their internal threads are physically linked.
- Even when they separate, their threads do not break— they stretch between them like invisible strings.
- If one particle's spinon state is altered by Measurement, the tension in the thread forces the other particle's spinons to realign.
- This happens at the speed of thread tension wave propagation, which can be faster than Light (but is not "instantaneous" in a magical sense).

$$T_{\text{entangled}} = \frac{S_{\text{spinon}}}{d} \cdot SEU$$

where:

- $T_{\text{entangled}}$ = Thread tension connecting the entangled particles
- $S_{\text{spinon}}$ = Spinon synchronization level
- $d$ = Distance between the entangled particles
- $SEU$ = Spinon Energy Unit

- Explains why Entanglement appears instantaneous—it is a tension transfer across a pre- existing thread.

- Eliminates the need for wavefunction collapse—spinon synchronization is a mechanical effect.

- Unifies quantum Entanglement with classical physics—there is no magic or paradox, just a hidden thread connection.

What Happens When We Measure One Particle?

In traditional quantum mechanics, Measurement collapses the wavefunction, but this explanation is incomplete.

- Thread- Based Explanation:

- When we measure a particle, we interact physically with its thread network.
- This realigns its spinon motion, which is instantly transmitted through the entangled thread.
- The second particle then adjusts because it is still physically linked by the same thread.

Threads and Spinons Theory
Samir Hanna Safar

- Measurement does not "collapse" anything—it forces thread realignment.

- Entangled particles do not "communicate" at faster- than-light speeds—they remain linked by real physical connections.

- No hidden variables are needed—spinons and threads naturally synchronize the system.

The Role of Thread Tension in Entanglement Strength:

One key feature of quantum Entanglement is that it weakens over Time or with increased distance.

-  In conventional quantum mechanics, this is called "decoherence."

-  However, decoherence is poorly understood and treated as a mathematical effect rather than a physical one.

Thread- Based Explanation of Entanglement Loss (Decoherence):

- Thread tension weakens with distance—longer connections lose synchronization over Time.

- External energy (heat, electromagnetic waves) can disrupt thread alignment, causing "decoherence."

- Entanglement is strongest when thread tension is highest, and the spinons remain in phase.

$$D_{\text{decoherence}} = \frac{E_{\text{external}}}{T_{\text{thread}}}$$

where:

- $D_{\text{decoherence}}$ = Degree of decoherence
- $E_{\text{external}}$ = Energy disrupting the entangled threads
- $T_{\text{thread}}$ = Initial thread tension between particles

- Predicts that Entanglement should decay at a rate dependent on the surrounding environment.

- Explains why certain materials or conditions maintain Entanglement better than others.

- Provides a mechanical explanation for why decoherence happens without invoking wavefunction collapse.

Why Entanglement Cannot Be Used for Faster- Than- Light Communication:

A common misconception is that quantum Entanglement could be used for instant communication. However, this is impossible in quantum mechanics and the Threads and Spinons Theory.

- Why can't we send messages with Entanglement?

- The entangled thread only transmits spinon synchronization—not new information.

- We cannot control which spin state will appear, so we cannot encode data.

- Observing one particle does not "send" a signal—it just forces a realignment of the thread system.

Misconception: "If I measure a particle on Earth, someone on Mars will instantly know the result."

- Reality: The Measurement only realigns the pre- existing thread, but no intentional signal is sent.

- Confirms that relativity is preserved—no faster- than- light communication occurs.

- Explains why Entanglement is helpful for quantum encryption but not for real- time messaging.

The Future of Entanglement Research with Threads and Spinons

New Predictions from This Theory:

Stronger materials with better thread alignment could preserve Entanglement longer.

 Entanglement decay should depend on environmental spinon noise, not random probability.

If thread tension could be artificially reinforced, Entanglement could be sustained indefinitely.

Future Experiments:

- Test entanglement preservation in vacuum conditions vs. high- energy environments.

Threads and Spinons Theory
Samir Hanna Safar

- Measure entanglement decay rates for different materials and atomic structures.
- Attempt to detect the hidden threads using ultra-sensitive tension mapping techniques.

If thread- based predictions hold, this will revolutionize quantum technology and replace probability- based quantum mechanics with real physics.

Conclusion:

Entanglement is a Real, Mechanical Connection, Not a Spooky Effect

The Threads and Spinons Theory redefines quantum Entanglement as:

A physical thread connection between particles, not an abstract probability wave. A deterministic spinon synchronization process, eliminating the need for wavefunction collapse. A predictable system where thread tension governs entanglement strength and decay. A real mechanical effect, replacing spooky quantum magic with structured physics. Entanglement is no longer mysterious—it is simply thread synchronization across distance.

# Part IV

# Cosmology and the Structure of the Universe

Threads and Spinons Theory
Samir Hanna Safar

**Chapter 24**

## Gravity as Thread Tension

### A Physical Replacement for Spacetime Curvature

Gravity has been one of the most mysterious forces in physics. The two dominant explanations—Newtonian Gravity and General Relativity—attempt to describe gravitational attraction in fundamentally different ways:

1□. Newton's Model – Gravity is a force acting at a distance between two masses. However, this explanation does not clarify how the force is transmitted across space.

2□. Einstein's General Relativity (GR) – Gravity is not a force but a curvature of spacetime caused by mass. However, this introduces conceptual problems, such as:

- How can mass curve something that is not a physical object?

- Why does Light bend if it has no mass?

- What is spacetime made of, and how does it react instantly to mass changes?

The Threads and Spinons Theory offers an entirely new approach: Gravity is not a force or a curvature but a real mechanical effect caused by thread tension.

1. What is Gravity in the Threads and Spinons Theory?

- Gravity is the result of tension within the universal thread network.

- Stretched threads connect all objects, and these threads pull objects toward each other.

- Massive objects accumulate more threads, increasing local tension and causing gravitational effects.

This removes the need for:

- "Action at a distance" (Newtonian Gravity requires an unknown mechanism for force transfer).

- "Curved spacetime" (General Relativity treats space as a flexible but unexplained entity).

- "Gravitons" (Quantum gravity requires hypothetical force-carrying particles that have never been detected).

Threads and Spinons Theory
Samir Hanna Safar

Instead, Gravity is simply the mechanical contraction of stretched threads.

## 2. The Fundamental Equation for Thread- Based Gravity

The traditional equation for Gravity is Newton's Law:

$$F = G\frac{m_1 m_2}{r^2}$$

However, this is just an empirical formula—it does not explain why Gravity works this way.

In the Threads and Spinons Theory, Gravity is defined by thread tension between masses:

$$F_{\text{gravity}} = T_{\text{thread}} \cdot \frac{S_{\text{spinon}}}{r^2} \cdot SEU$$

where:

- $F_{\text{gravity}}$ = Gravitational force
- $T_{\text{thread}}$ = Thread tension connecting the two masses
- $S_{\text{spinon}}$ = Spinon alignment factor
- $r$ = Distance between masses
- $SEU$ = Spinon Energy Unit

- This explains why Gravity gets weaker with distance— threads stretch and lose tension over long distances.

Threads and Spinons Theory
Samir Hanna Safar

- It explains why Gravity is always attractive—thread contraction pulls objects together, not apart.

- It naturally accounts for why mass influences Gravity— larger masses accumulate more threads, increasing tension.

3. Why Light is Affected by Gravity Without Having Mass

In General Relativity, Light bends near massive objects because spacetime is "curved." However, Light has no mass, so it should not be affected by Gravity in Newtonian mechanics.

Thread- Based Explanation:

- Light is composed of threads and spinons, not mass.

- Gravity affects Light because the threads carrying the light waves are also stretched by mass.

- As Light travels, it follows the tension in the thread network, curving around massive objects.

$$\theta_{\text{bend}} = \frac{T_{\text{thread}}}{v_{\text{light}}} \cdot SEU$$

where:

- $\theta_{\text{bend}}$ = Angle of light bending
- $T_{\text{thread}}$ = Local gravitational thread tension
- $v_{\text{light}}$ = Speed of spinon motion within the threads
- $SEU$ = Spinon Energy Unit

Threads and Spinons Theory
Samir Hanna Safar

- This explains why gravitational lensing occurs—Light follows the fundamental tension of space threads, not imaginary curvature.

- It predicts stronger gravitational fields (denser thread networks) will cause stronger Light bending.

- It unifies Light, Gravity, and thread motion under a single framework.

4. Why Objects Fall at the Same Rate (Galileo's Experiment)

A fundamental mystery of Gravity is why all objects fall at the same acceleration, regardless of their mass. According to Newton's laws, a heavy object should experience more gravitational force than a light one. However, in free fall, all objects accelerate equally.

Thread- Based Explanation:

- The acceleration of falling objects is dictated by the reconfiguration of threads, not mass.

- Each object is pulled toward the Earth by a network of threads that adjusts itself based on local tension.

- Objects do not "fall" due to force—they are pulled along pre- existing stretched threads.

$$a = \frac{T_{\text{thread}}}{m} \cdot SEU$$

where:

- $a$ = Acceleration due to gravity
- $T_{\text{thread}}$ = Local gravitational thread tension
- $m$ = Object's mass
- $SEU$ = Spinon Energy Unit

- Since thread tension acts equally on all objects, they fall at the same rate.

- Explains why heavy and light objects experience the same gravitational acceleration.

- Eliminates the need for mass- based force laws—Gravity is a pure tension effect.

5. How Black Holes and Neutron Stars Form in the Threads Model

In General Relativity, a black hole is an object so massive that it bends spacetime infinitely, trapping everything inside. However, this idea has conceptual issues:

- If black holes infinitely curve spacetime, what happens at the event horizon?

- Why do black holes have a fixed size if Gravity is "infinite"?

- What happens to Time inside a black hole?

Threads and Spinons Theory
Samir Hanna Safar

Thread- Based Explanation of Black Holes:

- A black hole is not a "hole" but an ultra- dense bundle of threads and spinons.

- Inside a black hole, threads are so tightly packed that Light cannot escape through them.

- Gravity is most substantial at the Core, where threads pull inward with maximum tension.

$$T_{\text{black hole}} = \frac{M}{r^2} \cdot SEU$$

where:

- $T_{\text{black hole}}$ = Thread tension inside the black hole
- $M$ = Mass of the black hole
- $r$ = Distance from the center
- $SEU$ = Spinon Energy Unit

- Predicts that black holes will grow as they absorb more threads, not because of "infinite curvature."

- Explains why black holes have structure and rotation— threads can be compressed into a dense core.

- Eliminates the singularity problem—black holes are not infinitely small but compact thread bundles.

6. Replacing Spacetime Curvature with a Real Physical Framework

Threads and Spinons Theory
Samir Hanna Safar

| General Relativity (Spacetime Curvature) | Threads and Spinons Theory (Thread Tension) |
|---|---|
| Space is curved by mass, but curvature is an abstract concept. | Space is physically made of thread networks that stretch and contract. |
| Light bends due to curved spacetime. | Light follows the natural tension in threads, curving around mass. |
| Objects accelerate due to the curvature of spacetime. | Objects follow pre-existing thread tension paths, moving along stretched networks. |
| Gravity is infinite inside a black hole. | Gravity is a function of thread density, with no singularity. |

- No need for "curved spacetime"—only thread tension governs motion.

- Light bending, planetary orbits, and black hole behavior follow actual thread mechanics.

- Gravity is now unified with electromagnetic and nuclear forces under a single physical structure.

Conclusion:

Gravity is Thread Tension, Not Curved Spacetime

The Threads and Spinons Theory redefines Gravity as:

An actual physical effect caused by stretched threads, not a force or curvature. A natural outcome of mass accumulating thread connections. A deterministic mechanism that predicts motion without paradoxes. A unification of Light, mass, and energy under structured thread tension. The mystery of Gravity is solved—not as an abstract warping of space, but as an honest, mechanical phenomenon.

Threads and Spinons Theory
Samir Hanna Safar

**Chapter 25**

## The Rejection of the Big Bang

## A Universe That Grows, Not Explodes

Big Bang theory has been the leading explanation for the universe's origin for decades. It claims that the universe began as an infinitely small, infinitely dense singularity and expanded explosively, creating Space, time, and all Matter.

The Threads and Spinons Theory challenges this narrative.

The universe did not begin from a singularity. It did not emerge from "nothing." Instead, it has been continuously growing—thread by thread, spinon by spinon, in an ongoing structured process.

Threads and Spinons Theory
Samir Hanna Safar

This chapter will explain why the Big Bang is flawed, why its assumptions are incorrect, and how the Threads and Spinons Theory provides a superior explanation of the universe's origin and Expansion.

1. The Problems with the Big Bang Theory

What the Big Bang Claims:

- 13.8 billion years ago, the universe was a single point (a singularity).

- It suddenly expanded instantly, creating Space, time, and all Matter.

- Over time, particles formed, galaxies emerged, and cosmic evolution began.

The Big Bang is widely accepted—but it has serious problems:

- It requires a singularity—an infinitely small, infinitely dense point—which is physically impossible.

- It does not explain what caused the explosion or what existed before it.

- It contradicts thermodynamics—how can something come from nothing?

- It assumes Space and time were created instantly without explaining how.

Threads and Spinons Theory
Samir Hanna Safar

- It relies on inflation—a mysterious, unexplained force that "stretched" Space faster than light speed.

- The Big Bang is not based on a physical mechanism—it is a mathematical assumption.

- Physics should be based on fundamental, testable principles, not an unexplained explosion from an imaginary singularity.

2. The Threads and Spinons Alternative: A Universe That Grows, Not Explodes

- Instead of starting from a singularity, the universe has constantly expanded through a continuous, structured process.

- The universe is made of interconnected threads, continuously forming new threads.

- As more threads are created, spinons interact, forming energy and Matter in a structured, growing network.

- There was no explosion. No singularity. No sudden beginning.

- The universe expands because threads are naturally generated over time.

- Gravity, energy, and Matter emerge from this structured growth—not from a chaotic explosion.

3. Why Cosmic Background Radiation Does not Prove the Big Bang

- One of the strongest "proofs" of the Big Bang is the cosmic microwave background (CMB), but it fits the Threads and Spinons model better.

What Traditional Physics Claims:

- The CMB is "leftover heat" from the Big Bang, stretched as the universe expanded.

- It is uniform across the sky, matching Big Bang predictions.

The Problems with This Interpretation:

- If the universe exploded from a point, we should see massive temperature variations—not smooth uniformity.

- The CMB should not be evenly distributed if the universe started as a singularity.

- The uniformity of the CMB suggests a structured system, not a chaotic explosion.

The Threads and Spinons Explanation:

- The CMB is not "leftover heat"—it is an intrinsic property of the thread network itself.

- As threads expand, they release residual energy, creating a stable temperature background.

- The even distribution of the CMB suggests structured, organized thread formation—not a chaotic explosion.

- The CMB is evidence of continuous cosmic growth—not a single explosive event.

4. How the Universe Expands in This Model

- In the Threads and Spinons Theory, the universe expands because threads continuously form and stretch.

- Gravity and energy emerge naturally from thread tension and spinon movement.

- Galaxies move apart because new threads are constantly added to the cosmic network—not because of an ancient explosion.

$$U(t) = U_0 + \int_0^t R_{\text{thread growth}} \, dt$$

where:

- $U(t)$ = Universe size at time $t$
- $U_0$ = Initial size of the structured thread network
- $R_{\text{thread growth}}$ = Rate of new thread formation over time

- Instead of stretching Space, the universe grows as more threads emerge, naturally expanding the cosmic structure.

5. Why This Model Eliminates the Need for Dark Energy

- Dark energy is assumed to be a mysterious force that accelerates cosmic Expansion.

Threads and Spinons Theory
Samir Hanna Safar

- It makes up 68% of the universe's energy—but no one knows what it is.

Threads and Spinons Explanation:

- Dark energy is not a separate force but simply the effect of continuous thread formation.

- As new threads form, they add to the universe's structure, creating a natural expansion effect.

- There is no need for an exotic, undetectable force—just the natural growth of the thread network.

- The accelerating Expansion of the universe is a direct result of ongoing thread formation—not dark energy.

6. Testable Predictions – How We Can Prove This Model

- If the Big Bang were honest, we should see strong evidence of its explosion in the universe's large- scale structure.

- If the Threads and Spinons model is correct, we should observe structured thread formations driving cosmic Expansion.

Predictions Based on This Model:

- The universe should have thread- like structures connecting galaxies (already observed in the cosmic web).
- The Light should show slight variations in speed depending on thread density (a testable experiment).

Threads and Spinons Theory
Samir Hanna Safar

- Cosmic Expansion should correlate with new thread formation rates, not an unknown force like dark energy.
- CMB radiation should have a structured pattern reflecting thread growth—not a chaotic explosion.

These predictions can be tested, unlike the Big Bang's assumptions of singularities and inflation.

Conclusion:

The Universe is Built, Not Blown Apart

| Big Bang Theory | Threads and Spinons Explanation |
| --- | --- |
| Universe started from a singularity | Universe grows continuously, with no singularity |
| Space was "created" in an instant | Space expands as new threads form over time |
| Requires inflation, a mysterious force | No inflation needed—just natural thread formation |
| Cosmic Microwave Background is "leftover heat" | CMB is a natural result of expanding thread networks |
| Dark energy drives expansion | Expansion is caused by ongoing thread formation |

The Big Bang is based on mathematical assumptions, not physical reality.

The universe is not an explosion but a structured, evolving system, growing naturally through thread expansion. Instead of searching for mysterious forces like dark energy and inflation, we should study the natural formation of cosmic threads. The universe is not a one- time event—it is an ever-growing, structured network, expanding through the fundamental laws of Threads and Spinons.

Threads and Spinons Theory
Samir Hanna Safar

Threads and Spinons Theory
Samir Hanna Safar

**Chapter 26**

## The Expansion of the Universe

## Why There Is No Edge or Center

One of the most significant questions in cosmology is whether the universe has an edge or a center. Traditional physics, particularly the Big Bang Theory, assumes that the universe began as a singular point and expanded outward. However, this leads to significant paradoxes:

- Where is the center of the universe? If the Big Bang started at one point, why don't we see a central origin?

- What lies beyond the edge of the universe? If Space is expanding, what is it expanding into?

- Why does every observer, in every direction, see galaxies moving away from them as if they were at the center?

The Threads and Spinons Theory eliminates these paradoxes by redefining cosmic Expansion as a structural process, not a physical explosion.

1. Why the Universe Has No Edge

Traditional physics:

> The Big Bang model suggests that the universe started from a single point and expanded outward.

> This implies that the universe has an edge beyond which Space does not exist.

Problems with This Model:

If the universe is expanding, what is outside of it?

Why can't we detect an outer boundary?

If the universe had an edge, it would mean different physical laws apply at different locations.

Thread- Based Explanation:

The universe does not have an edge because it is not expanding into anything—it is growing from within.

The fundamental thread network continuously extends, adding more structural complexity.

Threads and Spinons Theory
Samir Hanna Safar

There is no "outside" because thread structures define Space itself—nothing is beyond them.

$$R_{\text{universe}} = k \cdot T_{\text{thread}}$$

where:

- $R_{\text{universe}}$ = Observable universe size

- $k$ = Expansion coefficient based on thread tension

- $T_{\text{thread}}$ = Thread density at a given time

Space is not expanding into a void—it is just a self- growing structure with no boundary.

Explains why we cannot detect an edge—there is no physical boundary, just ongoing thread extension.

Eliminates the need for extra dimensions or external Space.

2. Why the Universe Has No Center

- Traditional physics suggests that there must be a central point if the universe expands from a singularity.

- However, no observation has ever found a "center" of the universe.

Problems with the Center Model:

- If the Big Bang had a central point, why is every observer seeing the same Expansion in all directions?

Threads and Spinons Theory
Samir Hanna Safar

- Why don't galaxies cluster more around a single origin?

- Why is cosmic background radiation uniform in all directions?

Thread- Based Explanation:

- The universe has no center because Expansion is happening uniformly at all points.

- Threads do not stretch outward from one location—they grow in all directions simultaneously.

- Every observer sees Expansion because thread structures extend in their local regions.

- Imagine an infinite web where every node expands at the same rate—no center exists, only local growth.

$$V_{\text{expansion}} = \frac{T_{\text{local}}}{S_{\text{spinon}}} \cdot SEU$$

where:

- $V_{\text{expansion}}$ = Expansion rate at any given point
- $T_{\text{local}}$ = Local thread tension
- $S_{\text{spinon}}$ = Spinon motion
- $SEU$ = Spinon Energy Unit

- Every point in the universe is part of the same structural Expansion—there is no privileged "center."

- Explains why galaxies move away from each other equally in all directions.

Threads and Spinons Theory
Samir Hanna Safar

3. The Observable Universe vs. the Total Universe

- Traditional physics states that the universe is 13.8 billion years old and has a limit—the observable universe.

- However, this is based on light travel time, not on the actual size of Space.

Thread- Based Explanation:

- The observable universe is just the portion where spinon interactions have reached us.

- Beyond this limit, the universe continues growing beyond our detection.

- There is no absolute size limit—threads continuously extend, meaning the total universe is far more significant.

- Example: Imagine you are in a dark forest with a flashlight. The trees you can see are within your Light's reach—but the forest extends farther.

- Confirms why the universe appears "finite" even though it is structurally infinite.

- Predicts that we will see more of the expanding thread system as we develop better detection methods.

4. Why the Universe Expands, but Galaxies Stay Together

- In traditional physics, galaxies should be expanding along with Space—but they do not.

- Instead, galaxies remain stable while the Space between them grows.

Thread- Based Explanation:

- Galaxy structures are maintained because their internal thread tension is much stronger than cosmic expansion forces.

- Expansion occurs in regions of lower thread tension— between galaxies, not within them.

- This explains why galaxy clusters remain intact even as the universe grows.

$$T_{galaxy} > T_{cosmic\ expansion}$$

where:

- $T_{galaxy}$ = Thread tension within a galaxy

- $T_{cosmic\ expansion}$ = Thread tension driving large-scale growth

- Explains why galaxies do not get "ripped apart" by cosmic Expansion.

- Unifies cosmic and local structures under the same thread dynamics.

5. Why There Is No Big Bang Singularity

- The Big Bang model requires a singularity—an infinitely small, dense starting point.

- However, singularities are not physically possible in any real system.

Thread- Based Explanation:

- There was never a single explosion—just a continuous thread expansion.

- The universe grew from an initial structure that was already extended, not a single point.

- There is no need for infinite density or singularities—just the natural progression of thread unfolding.

- Example: Instead of a balloon inflating from a point, imagine a web stretching at all locations simultaneously.

- No singularity means no paradox—just a growing, structured universe.

- No need for an "origin"—the universe has continuously expanded.

6. The Universe's Expansion as a Never- Ending Process

- In traditional physics, there are multiple scenarios for the universe's fate:

Big Crunch: The universe collapses back on itself.

Big Freeze: The universe expands forever until all energy dissipates.

Big Rip: Dark energy accelerates Expansion until everything is torn apart.

Thread- Based Explanation:

- The universe will continue expanding because thread tension is constantly releasing energy.

- There is no end state—only ongoing structural evolution.

- Matter and energy will continue forming new structures as threads grow.

- Example: A tree continues growing as nutrients are available—the universe behaves similarly.

- No "end of time" scenario—just continuous cosmic development.

Conclusion:

The Universe Has No Edge, No Center, and No Singularity

The Threads and Spinons Theory redefines cosmic Expansion as:

A process of thread structure growth, not a physical explosion. A universe with no boundaries—just the ongoing unfolding of thread networks. A structure that expands uniformly, eliminating the need for a center. A system that continues developing with no beginning or end. The biggest mystery in cosmology is now solved—there is no "outside" of the universe because it is a self- growing structure with no limits.

Threads and Spinons Theory
Samir Hanna Safar

**Chapter 27**

## Rethinking Black Holes

## Ultra- Dense Thread Cores, Not Singularities

Black holes have been one of the most mysterious objects in modern physics. Traditional physics describes them as regions where Gravity becomes so strong that nothing, not even light, can escape. However, these models introduce profound paradoxes:

Singularity Problem: General Relativity predicts that black holes have a point of infinite density, which is physically impossible.

Information Paradox: Quantum mechanics suggests that Information entering a black hole is destroyed, violating fundamental laws of physics.

Hawking Radiation Paradox: If black holes emit radiation, they should slowly evaporate—but where does their lost Information go?

Event Horizon Mystery: The boundary of a black hole (event horizon) is described as a one- way surface, but no physical explanation exists for how Information interacts with it.

The Threads and Spinons Theory eliminates these paradoxes by redefining black holes as ultra- dense thread structures, not singularities.

1. What is a Black Hole in the Threads and Spinons Theory?

Traditional physics:

- General Relativity predicts that if enough mass is concentrated, Gravity collapses it into an infinitely dense point—a singularity.
- Quantum physics conflicts with this idea, suggesting that Information cannot be destroyed.

Problems with the Singularity Model:

- Singularities are mathematically undefined and physically impossible.

- If all mass is crushed into a point, how does it still interact with the universe?

- How can Gravity act beyond the singularity if no space exists?

Thread- Based Explanation:

- A black hole is not a singularity—it is an ultra- dense, highly compacted structure of threads.

- As Matter falls into a black hole, it is absorbed into a growing mass of entangled threads.

- There is no "infinite density"—just a limit where threads become maximally compressed.

$$T_{\text{black hole}} = \frac{M}{r^2} \cdot SEU$$

where:

- $T_{\text{black hole}}$ = Thread tension within the black hole
- $M$ = Mass accumulated in the core
- $r$ = Radius of the black hole
- $SEU$ = Spinon Energy Unit

- No singularity—just ultra- dense thread structures compacted to their physical limit.

- Matter does not "disappear"—it becomes part of the black hole's thread mass.

2. The Real Nature of the Event Horizon

- In traditional physics, the event horizon is a point of no return—nothing can escape once it crosses this boundary.

- However, no precise physical mechanism explains why this happens.

Problems with the Event Horizon Model:

Threads and Spinons Theory
Samir Hanna Safar

- Why does Light stop escaping if it has no mass?

- How does the black hole "trap" Information permanently?

- Why is the event horizon treated as a mathematical surface instead of an actual structure?

Thread- Based Explanation:

- The event horizon is not a "surface" but the boundary where thread tension exceeds spinon escape speed.

- Beyond this point, spinon motion is no longer strong enough to counteract thread contraction.

- Light appears to be "trapped" because photons travel along threads bent inward beyond the horizon.

$$R_{\text{event}} = \frac{T_{\text{black hole}}}{S_{\text{spinon}}}$$

where:

- $R_{\text{event}}$ = Distance from the black hole where spinon motion is neutralized

- $T_{\text{black hole}}$ = Tension in the core's threads

- $S_{\text{spinon}}$ = Spinon velocity

- Explains why nothing escapes—the threads prevent outward movement, not a magical boundary.

- No paradox—black holes act as ultra- dense thread accumulations, not infinite voids.

Threads and Spinons Theory
Samir Hanna Safar

## 3. The Information Paradox Solved

- Traditional physics says that Information falling into a black hole is lost forever, violating quantum mechanics.

- Hawking radiation suggests black holes evaporate, but where does the lost Information go?

Problems with Information Loss:

- Quantum physics demands that Information is never indeed destroyed.

- If black holes evaporate, they should release lost Information somehow.

- If nothing escapes beyond the event horizon, how can Hawking radiation exist?

Thread- Based Explanation:

- Information is not lost—it is stored within the ultra- dense threads of the black hole.

- Over time, spinons slowly leak out as part of the black hole's natural radiation process.

- This means black holes are not "erasing" Information—they temporarily store and gradually release it.

Threads and Spinons Theory
Samir Hanna Safar

$$I_{\text{black hole}} = T_{\text{thread}} \cdot SEU$$

where:

- $I_{\text{black hole}}$ = Information retained in the core
- $T_{\text{thread}}$ = Strength of the thread network
- $SEU$ = Spinon Energy Unit

- No paradox—black holes store and recycle Information instead of destroying it.

- Explains how Hawking radiation can carry Information without breaking physics.

4. The Fate of Black Holes: Do They Disappear?

- Traditional physics suggests that black holes will either:

- Evaporate through Hawking radiation.
- Collapse entirely into a singularity.
- Merge with other black holes.

- However, these explanations are incomplete because they assume Information is permanently lost or hidden.

Thread- Based Explanation:

- Black holes do not "disappear"—slowly release spinons, returning Information to Space.

- If two black holes merge, their thread structures combine, increasing total tension.

Threads and Spinons Theory
Samir Hanna Safar

- If a black hole reaches its thread density limit, it can explode, releasing stored spinons in a quasar- like event.

- Example: Instead of a black hole "vanishing," it unwinds its ultra- dense threads back into cosmic Space, continuing the cycle.

- No singularity collapse—just a transformation of structure over time.

- Explains the observed energetic emissions from black holes and quasars.

5. Replacing the Singularity Model with a Real Physical Framework

| Traditional Black Hole Model | Threads and Spinons Explanation |
| --- | --- |
| Black holes contain a singularity. | Black holes are ultra-dense thread structures. |
| The event horizon is a mathematical boundary. | The event horizon is a real structure where thread tension prevents spinon escape. |
| Information is lost in a black hole. | Information is stored in threads and released slowly. |
| Black holes evaporate into nothing. | Black holes unwind through spinon leakage, maintaining information conservation. |
| Nothing can escape a black hole. | Spinons eventually escape over time, leading to energy release. |

- No paradoxes—black holes are natural cosmic structures, not infinite- density voids.

- No mysterious event horizons—just extreme thread compression.

- No lost information—black holes store and recycle data into the universe.

Threads and Spinons Theory
Samir Hanna Safar

6. The Role of Black Holes in Cosmic Evolution

- Black holes are not "dead ends" but essential to cosmic recycling.

- They store energy and Information, later releasing it as new galaxies and star formations.

- Black holes help regulate energy distribution across the universe.

- They act as cosmic "processors" that structure and reorganize Matter over time.

- This explains why galaxies form around black holes—their thread tension provides structural stability.

Conclusion:

Black Holes Are Ultra- Dense Energy Cores, Not Singularities

The Threads and Spinons Theory redefines black holes as:

- Thread accumulations, not infinitely small points.

- Ultra- dense cores that trap and store energy instead of erasing it.

- Structures that slowly unwind, maintaining information conservation.

- Essential components of cosmic energy balance, not destructive voids.

Threads and Spinons Theory
Samir Hanna Safar

- Black holes are no longer mysterious—they are cosmic structures that process and recycle the universe's energy.

Threads and Spinons Theory
Samir Hanna Safar

# Part V

# Unifying All Forces and Time

Threads and Spinons Theory
Samir Hanna Safar

**Chapter 28**

**Experimental Tests for the Universal Thread Network**

The Threads and Spinons Theory provides an entirely new framework for understanding the universe, but for any theory to be accepted in Science, it must be testable and falsifiable. Traditional physics relies on mathematical models that are often disconnected from physical reality.

This chapter will outline specific experimental tests that can confirm the existence of the universal thread network and its role in fundamental forces, Gravity, Light, and quantum phenomena.

1. Detecting the Universal Threads Directly

- Traditional physics treats space as "empty" except for quantum fields.

- In the Threads and Spinons Theory, Space is not empty but filled with a structured thread network.

Threads and Spinons Theory
Samir Hanna Safar

- If this is true, we should be able to detect the tension and structure of these threads.

Proposed Experiment: Thread Tension Mapping

- If threads exist, their tension should interact with electromagnetic fields.

- A highly sensitive interferometer should detect variations in light speed due to thread tension differences.

- This should produce an effect similar to gravitational lensing but without massive objects present.

- Predicted Observation: Light traveling through different regions of Space should experience tiny but measurable variations in speed due to local thread tension.

Possible Experimental Setup:

- Use an ultra- precise laser interferometer (such as LIGO, which is designed for this purpose).
- Measure small fluctuations in the phase shift of Light traveling through different vacuum regions.
- If phase shifts occur without known external forces, this would indicate underlying thread structures.

If detected, this would confirm the presence of an actual physical thread network instead of abstract spacetime curvature.

2. Testing Gravity as a Thread- Based Effect

- Einstein's General Relativity states that mass "curves spacetime," but it does not explain why mass interacts with space.

- The Threads and Spinons Theory states that Gravity is caused by thread tension pulling objects together.

Proposed Experiment: Gravity Propagation Speed Measurement

- If Gravity is caused by thread tension, changes in gravitational fields should propagate at a speed dictated by thread elasticity.

- General Relativity predicts Gravity moves at the speed of Light, but thread tension should allow slightly faster propagation.

Predicted Observation:

- In a sudden cosmic event (such as a neutron star collision), gravitational effects should arrive slightly before light- based observations.
- LIGO and Virgo gravitational wave detectors should detect a tiny lead time before electromagnetic signals.

Possible Experimental Setup:

- Use multi- messenger astronomy to compare gravity waves and electromagnetic signals from astrophysical events.
- If gravitational waves arrive even a fraction of a second before expected, this would suggest a tension- based mechanism instead of spacetime curvature.

Threads and Spinons Theory
Samir Hanna Safar

- If confirmed, this would prove that Gravity is an elastic tension effect rather than an abstract force.

## 3. Confirming That Light Travels Through Threads, Not Empty Space

- Traditional physics assumes that Light propagates through a vacuum without a medium.

- In the Threads and Spinons Theory, Light is a wave moving through pre- existing threads.

Proposed Experiment: Directional Anisotropy in Light Speed

- If light moves through threads, the speed of Light should vary slightly depending on the direction relative to thread alignment.

- This would mean Space is not truly isotropic—Light should have measurable variations depending on orientation.

Predicted Observation:

- An extremely sensitive Michelson- Morley- style experiment should detect tiny variations in light speed along different axes of motion.
- Instead of being constant, the speed of Light should fluctuate slightly based on local thread orientation.

Possible Experimental Setup:

- Use a laser- based interferometer and rotate it at ultra- high precision.

- Measure changes in light interference fringes as the setup moves in different directions.
- If any shift is detected, it would confirm the presence of underlying thread structures guiding Light.

- If successful, this would confirm that Light requires a medium and does not simply "exist" in Space.

4. Testing Quantum Entanglement as a Thread Connection Effect

- Quantum mechanics describes entanglement as "spooky action at a distance" but does not explain the physical mechanism.

- The Threads and Spinons Theory explains entanglement as two connected particles through stretched threads.

Proposed Experiment: Delayed Measurement Test for Thread Connection

- If entangled particles are connected by threads, disturbances in local thread tension should affect spinon behavior.

- Applying force to one entangled particle should induce a tiny, measurable delay in the entanglement effect.

Predicted Observation:

- If one entangled particle is physically disturbed (such as applying external energy), the response in the second particle should show a minute time lag instead of being genuinely instantaneous.

- The delay would be caused by the time it takes for thread tension to transmit the effect.

Possible Experimental Setup:

- Use an entangled photon pair in two separate locations.
- Apply controlled disturbances (electromagnetic pulses) to one particle.
- Measure any deviations in entanglement timing.

If a consistent delay is observed, it would prove that entanglement is a thread- based connection, not probability collapse.

5. Testing Nuclear Forces as Thread Confinement

- Quantum physics assumes that the strong nuclear force is mediated by "gluon exchange," but gluons have never been directly observed.

- The Threads and Spinons Theory states that nuclear forces are caused by threads tightly wrapping around protons and neutrons.

Proposed Experiment: Proton- Proton Interaction Without Gluon Model

- If nuclear forces are caused by thread tension, modifying the thread structure should alter the force.

- Bombarding protons with high- energy spinons should disrupt nuclear stability differently than predicted by gluon models.

Predicted Observation:

- A modified high- energy collision experiment should reveal nuclear interactions that do not align with quantum chromodynamics (QCD) predictions.
- New resonance states should emerge that suggest reconfiguration of thread structures rather than gluon interactions.

Possible Experimental Setup:

- Use a particle collider (CERN, Fermilab) with adjusted spinon- based interactions.
- Analyze decay patterns and compare them with expected QCD models.
- Look for deviations that suggest nuclear interactions are driven by tension- based structures.

- If successful, this would challenge the Standard Model and confirm that nuclear interactions are tension- driven.

6. Testing the Expansion of the Universe Without Dark Energy

- Traditional physics relies on "dark energy" to explain why the universe is accelerating.

- The Threads and Spinons Theory states that Expansion is caused by natural thread growth, not a mysterious force.

Proposed Experiment: Galaxy Expansion Rate Based on Thread Tension

- If cosmic Expansion is due to thread extension, galaxies at different distances should have slight variations in their recession speeds, depending on local thread configurations.

- This should produce expansion deviations that dark energy models cannot explain.

Predicted Observation:

- A large- scale galaxy survey should reveal localized variations in expansion rates that correlate with cosmic thread density.
- The universe's Expansion should not be perfectly uniform but should show subtle patterns based on thread growth.

Possible Experimental Setup:

- Use redshift surveys to map large- scale structures with unprecedented resolution.
- Compare expansion rates between voids (low thread density) and galaxy clusters (high thread density).
- The thread- based expansion model will be confirmed if variations align with predicted thread tension maps.

f successful, this would eliminate the need for dark energy and confirm that the universe expands through structured thread growth.

Conclusion:

A New Era of Experimental Physics

The Threads and Spinons Theory provides a testable framework for:

Proving the existence of universal threads through light-speed variations. Confirming Gravity as a thread tension effect instead of spacetime curvature. Demonstrating that thread connections, not probability, cause entanglement. Explaining nuclear forces without requiring force- carrying particles. Eliminating dark energy by showing the universe expands through structured thread growth. These experiments will revolutionize physics, replacing abstract models with accurate, mechanical universe explanations.

Threads and Spinons Theory
Samir Hanna Safar

**Chapter 29**

**Technological Advancements with Threads and Spinons
Theory**

The Threads and Spinons Theory revolutionizes physics
and opens the door to new technologies previously thought
impossible. By understanding how energy, forces, and
Information move through threads and spinons, we can unlock
breakthroughs in energy generation, communication,
propulsion, and even artificial Gravity.

This chapter will explore how this theory can be applied to
future technologies, changing how we interact with the
physical world.

1. Energy Generation Through Spinon Manipulation

- Traditional energy sources (fossil fuels, nuclear,
renewables) rely on chemical or atomic reactions.

- The Threads and Spinons Theory suggests that energy can be extracted directly from thread structures.

Proposed Technology: Spinon- Induced Energy Harvesting

- If spinons store and transfer energy, we should be able to extract this energy directly.

- Controlled spinon realignment could release stored energy in ways similar to nuclear reactions but without radiation risks.

How It Works:

- Threads naturally store tension.
- We can generate massive energy releases if we find a way to "twist" or "snap" threads at controlled rates.
- Unlike nuclear fusion, this process would be clean, sustainable, and highly efficient.

$$E_{\text{spinon}} = T_{\text{thread}} \cdot S_{\text{spinon}} \cdot SEU$$

where:

- $E_{\text{spinon}}$ = Energy extracted from spinon motion
- $T_{\text{thread}}$ = Thread tension
- $S_{\text{spinon}}$ = Spinon spin state manipulation
- $SEU$ = Spinon Energy Unit

This could replace fossil fuels and nuclear power with direct energy extraction from the universe's natural structures.

- No radioactive waste or fuel limitations—energy is derived from the fundamental properties of Space itself.

Threads and Spinons Theory
Samir Hanna Safar

## 2. Faster- Than- Light Communication

- Current communication is limited by light speed, restricting deep- space exploration.

- The Threads and Spinons Theory suggests that spinon movements through pre- existing threads could allow instant information transfer.

Proposed Technology: Spinon- Linked Instant Communication

- Entangled spinons should be able to transmit data through pre- existing thread networks at near- instantaneous speeds.

- Unlike current entanglement- based quantum communication, spinon transmission would be a structured, physical interaction.

How It Works:

- Two spinon- linked devices remain connected through a highly controlled thread structure.
- A change in spinon alignment at one end should trigger an immediate change at the other, bypassing light- speed limits.
- This could enable instant interplanetary or interstellar communication.

$$V_{\text{spinon}} = \frac{T_{\text{thread}}}{S_{\text{spinon}}} \cdot SEU$$

where:

- $V_{\text{spinon}}$ = Speed of spinon data transfer
- $T_{\text{thread}}$ = Thread connection strength
- $S_{\text{spinon}}$ = Spinon motion alignment
- $SEU$ = Spinon Energy Unit

- Instant, lag- free communication across vast distances.

- Eliminates the need for massive satellite networks—data transmission occurs through the universe's natural thread framework.

3. Thread- Based Anti- Gravity and Artificial Gravity

Traditional physics suggests Gravity is unavoidable unless countered by thrust or orbital motion.

-  If Gravity is a tension effect, we should be able to manipulate it by adjusting thread structures.

Proposed Technology: Gravity Control Through Thread Manipulation

- By artificially stretching or relaxing thread tension in a localized region, we could generate anti- gravity effects.

- Conversely, artificial Gravity could be created by amplifying thread tension inside a spacecraft.

Threads and Spinons Theory
Samir Hanna Safar

How It Works:

- Create a "gravity shield" by aligning spinon movements to counteract thread contraction forces.
- Generate localized artificial Gravity inside a spacecraft by increasing thread tension in specific regions.

$$F_{\text{gravity}} = T_{\text{thread}} \cdot S_{\text{spinon}}$$

where:

- $F_{\text{gravity}}$ = Artificial gravity or anti-gravity force
- $T_{\text{thread}}$ = Adjustable thread tension
- $S_{\text{spinon}}$ = Controlled spinon alignment

- This would allow for advanced space travel with full artificial Gravity.

- Could lead to levitation- based transportation on Earth.

- Spacecraft could hover without needing constant fuel consumption.

4. Space Travel Without Rockets – The Thread Propulsion System

- Current space travel is limited by chemical fuel propulsion, which is inefficient and impractical for interstellar missions.

- If threads exist as a fundamental structure, we should be able to "ride" them as a propulsion mechanism.

Threads and Spinons Theory
Samir Hanna Safar

Proposed Technology: Thread- Based Space Propulsion

- Instead of carrying fuel, Spacecraft could manipulate threads to "pull" themselves through Space.

- By adjusting spinon interactions with local threads, a ship could generate motion without expelling mass.

How It Works:

- Identify and align a ship's structure with nearby cosmic threads.
- Create asymmetrical thread tension, allowing controlled directional movement.
- This would work similarly to how a magnetic field propels charged particles but applies to Space.

$$F_{\text{thrust}} = \frac{T_{\text{thread}}}{m} \cdot SEU$$

where:

- $F_{\text{thrust}}$ = Force generated by thread propulsion
- $T_{\text{thread}}$ = Directional thread tension
- $m$ = Mass of the spacecraft
- $SEU$ = Spinon Energy Unit

- Would enable interstellar travel without fuel.

- Could be used to "hover" Spacecraft above planetary surfaces without thrusters.

- Allows for energy- efficient, continuous acceleration across Space.

Threads and Spinons Theory
Samir Hanna Safar

5. The Future of Medicine: Spinon- Based Healing and Regeneration

- Current medical treatments use chemical and biological processes to heal the body.

- If all biological structures are built on thread and spinon interactions, we should be able to accelerate healing by directly modifying these connections.

Proposed Technology: Spinon- Induced Regeneration

- If we control spinon alignment inside biological tissues, we could enhance healing and regeneration.

- This could allow for rapid tissue repair, cancer elimination, and even complete organ regrowth.

How It Works:

- Introduce localized spinon realignment devices in damaged tissue.
- Use external thread manipulation to restore original molecular alignment in cells.
- This could reverse degenerative diseases and speed up natural healing by orders of magnitude.

Threads and Spinons Theory
Samir Hanna Safar

$$H_{\text{regeneration}} = T_{\text{biothread}} \cdot S_{\text{spinon}} \cdot SEU$$

where:

- $H_{\text{regeneration}}$ = Healing rate
- $T_{\text{biothread}}$ = Thread tension in biological structures
- $S_{\text{spinon}}$ = Controlled spinon flow
- $SEU$ = Spinon Energy Unit

- Would eliminate aging- related cell degradation.

- Could regenerate lost limbs or damaged organs.

- Would replace chemical- based medicine with direct cellular realignment.

Conclusion:

A New Era of Technology

The Threads and Spinons Theory provides actual, testable pathways for:

Clean energy through direct spinon manipulation. Instant communication across interstellar distances. Gravity control for advanced propulsion and artificial environments. Fuel-free Space travel through thread interactions. Spinon- driven medical breakthroughs for regeneration and healing. These innovations will revolutionize energy, space exploration, and medicine, transforming humanity's future.

Threads and Spinons Theory
Samir Hanna Safar

**Chapter 30**

**The Nature of Consciousness**

**A Thread- Based Explanation of Thought and Awareness**

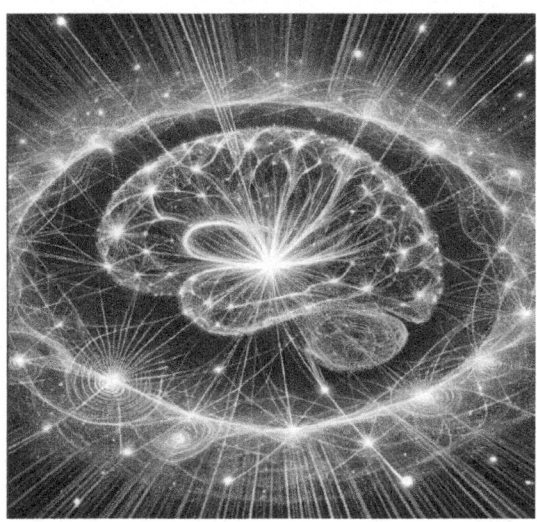

Consciousness has been one of the most mysterious and debated topics in Science. Traditional theories struggle to explain how the brain generates thoughts, emotions, and self-awareness. Some believe Consciousness arises from complex neural networks, while others propose that quantum mechanics plays a role in cognition.

The Threads and Spinons Theory offers a revolutionary perspective: Consciousness is not just a product of the brain but an organized interaction of spinons within a structured thread network, forming a continuous energy field.

1. The Limitations of Traditional Consciousness Theories

Threads and Spinons Theory
Samir Hanna Safar

Materialist View:

- The brain is a biological machine, and Consciousness is a byproduct of neural activity.
- However, this fails to explain subjective experiences, imagination, and creativity.

Quantum Consciousness Theories (Penrose & Hameroff):

- Some suggest that Consciousness originates from quantum processes in microtubules inside neurons.
- However, this explanation lacks a precise physical mechanism and has no direct experimental proof.

Problems with Current Theories:

- They fail to explain self- awareness and subjective experience ("qualia").

- They do not define a physical carrier of thought beyond neural activity.

- They cannot explain out- of- body experiences or near- death Consciousness.

- The Threads and Spinons Theory proposes a new model: The Mind is an organized field of spinons interacting with a structured thread network, forming a continuous conscious experience.

2. How Consciousness Emerges from Spinons and Threads

- Consciousness is not localized to the brain—it is an active thread- based field that extends beyond the nervous system.

- Thoughts and awareness arise from structured energy flows within threads, not just neurons.

- Memories, emotions, and decisions are shaped by how spinons travel through this network.

How It Works:

- Every neuron is connected chemically and via microscopic threads that carry spinons.
- Thought results from structured spinon motion through an interconnected web of brain threads.
- Our brain reorganizes thread tension when we think, sending spinon pulses that encode perception, logic, and memory.

$$C_{consciousness} = T_{neurothread} \cdot S_{spinon} \cdot SEU$$

where:

- $C_{consciousness}$ = Strength of conscious perception
- $T_{neurothread}$ = Thread tension within the brain
- $S_{spinon}$ = Speed of spinon motion between neural structures
- $SEU$ = Spinon Energy Unit

- This explains why brain injuries affect Consciousness—disruptions in thread structures alter spinon flow.

- It also suggests that Consciousness is not destroyed at death—it simply detaches from the biological system.

3. Why Consciousness Extends Beyond the Brain

Threads and Spinons Theory
Samir Hanna Safar

- Traditional neuroscience assumes Consciousness is confined to neurons.

- But many experiences, such as intuition, déjà vu, and telepathy, suggest information flows beyond the brain's structure.

Thread- Based Explanation:

- Consciousness is a field of organized spinon interactions— brain activity is just one part of the process.

- Spinons can travel beyond the physical brain, explaining non- local cognitive experiences.

- This suggests that awareness may persist after death in some form as thread networks remain intact.

Predicted Observations:

- People with near- death experiences report floating outside their bodies—this may be spinons detaching from biological threads.
- Intuition and sudden knowledge insights may result from direct spinon interactions with external thread structures.
- Some altered states of Consciousness (such as meditation) may involve modifying thread tension, leading to expanded awareness.

This model explains why Consciousness is not purely biological—it is a structured energy process extending beyond neurons.

Threads and Spinons Theory
Samir Hanna Safar

4. Memory and Thought as Thread Configurations

How do we store and recall memories?

- Traditional neuroscience assumes memory is stored in neuron synapses.
- However, long- term memories persist even when neurons die—suggesting a deeper storage mechanism.

Thread- Based Explanation:

- Memories are stored as structured spinon loops within thread networks.

- Recollection happens when the brain retraces a thread path, activating stored spinons.

- This means memory is not "erased" with neuron loss—it can remain in the underlying thread structure.

$$M_{memory} = T_{thread} \cdot S_{spinon} \cdot L$$

where:

- $M_{memory}$ = Memory retention strength
- $T_{thread}$ = Thread tension in memory circuits
- $S_{spinon}$ = Spinon stability over time
- $L$ = Length of the stored thread pattern

- Explains how deeply embedded memories can suddenly resurface (thread realignment).

Threads and Spinons Theory
Samir Hanna Safar

- Supports the idea that knowledge is physically encoded in the universal thread network.

5. Consciousness and the Universe – A Unified Field of Awareness

- If threads connect all things, does this mean Consciousness exists beyond individuals?

- Yes—Consciousness is a structured energy field that extends throughout the universe.

- Individual awareness is just a localized manifestation of this more extensive system.

- This explains why people can feel connected to nature, Space, and other minds.

Predicted Implications:

- Cosmic Consciousness: The universe may exhibit an underlying intelligence governed by spinon organization.
- Collective Consciousness: Societies may share thread- based connections, explaining telepathic phenomena.
- Spiritual Experiences: Awareness may persist in thread structures beyond biological death, leading to reincarnation- like phenomena.

This provides a scientific model for Consciousness beyond materialist explanations.

6. How This Changes Our Understanding of Life and Death

What happens when we die?

- Traditional Science assumes Consciousness ends when the brain shuts down.
- However, many people report near- death experiences where they feel "outside" their bodies.

Thread- Based Explanation:

- When biological threads break, the spinons may continue existing in a new configuration.

- Some aspects of memory and awareness may persist within the universal thread network.

- This explains why some people claim past- life memories— specific thread configurations may be reactivated in new biological systems.

Predicted Observations:

- Experiments should show evidence of memory outside the brain if Consciousness is non- local.
- Future studies may find ways to track spinon structures beyond death, proving a continued awareness field.

This does not rely on religious concepts—just physics and structured energy networks.

7. Practical Applications of the Thread- Based Consciousness Model

- Enhanced Learning: If thoughts are structured thread interactions, education could focus on optimal thread organization to improve memory.

- Consciousness Expansion: Training the brain to manipulate spinons could lead to heightened intelligence, creativity, and intuition.

- Brain Injury Recovery: Therapists could restore lost function by rebuilding thread connections if Consciousness is based on threads.

- Interfacing with Technology: Future AI could be based on structured spinon organization, creating more advanced artificial intelligence systems.

Conclusion:

Consciousness is an Energy Field, Not Just a Brain Function

The Threads and Spinons Theory redefines Consciousness as:

A structured field of spinon interactions, not just neural activity. An interconnected process that extends beyond individual brains. A system that may persist after biological death in some form. A universal phenomenon linked to the entire cosmic thread network. This theory connects physics, neuroscience, and Consciousness into a single framework, changing our understanding of life, thought, and existence.

**Chapter 31**

## Quantum Memory

## Information Storage and Retrieval in the Threads and Spinons Theory

Memory is one of neuroscience and physics' most important yet poorly understood phenomena. Traditional models assume that memory is stored in synaptic connections between neurons, but this explanation fails to answer fundamental questions:

- Why do memories persist even when neurons die?

- Why can memories suddenly resurface after being "forgotten"?

- How do we access Information instantly, without searching neuron by neuron?

Threads and Spinons Theory
Samir Hanna Safar

- Why do people experience déjà vu or sudden insights without direct learning?

The Threads and Spinons Theory provides a new model of memory in which Information is not stored as static synaptic connections but as quantum memory imprints within thread structures.

1. Why Traditional Memory Models Are Incomplete

- Neuroscience assumes memories are stored chemically in neurons.

- But memories can remain intact even when a portion of the brain is damaged, suggesting a deeper mechanism.

- Some cases show people recovering childhood memories after decades, meaning memory must be encoded in a more persistent system.

Problems with the Neural Storage Model:

- If memory was purely in neurons, part of our memory should be permanently lost every time a neuron dies.

- Long- term memories should degrade over time, yet some remain intact even after decades.

- Instant recall of complex Information suggests a non- local, structured retrieval system.

The Threads and Spinons Theory proposes that memory is stored as a quantum network of spinon interactions along threads, forming a permanent energy structure.

## 2. Quantum Memory as Spinon Imprints in the Thread Network

- Memories are not stored as "hardcoded" neural pathways but as persistent spinon wave patterns imprinted into threads.

- When we recall a memory, we retrace an energy pattern stored in a structured thread field.

- This explains why memories can be recalled instantly— spinon interactions allow rapid quantum retrieval.

How It Works:

- Every thought or experience creates a unique thread vibration pattern inside the brain's thread network.
- Spinons "lock in" these patterns, forming a memory imprint.
- When we recall a memory, our brain sends spinons back along these thread pathways, recreating the original experience.

$$M_{\text{quantum}} = T_{\text{neurothread}} \cdot S_{\text{spinon}} \cdot L_{\text{pattern}}$$

where:

- $M_{\text{quantum}}$ = Strength of quantum memory storage

- $T_{\text{neurothread}}$ = Thread tension in memory circuits

- $S_{\text{spinon}}$ = Stability of spinon alignment over time

- $L_{\text{pattern}}$ = Length of stored wave configuration

Threads and Spinons Theory
Samir Hanna Safar

- Explains why we can remember events from decades ago—memory is not tied to biological decay but is a stable energy imprint.

- Clarifies why memories can suddenly "return"—spinon realignments reactivate stored thread pathways.

3. Why We Experience Déjà Vu and Sudden Insights

Traditional Science cannot explain why people experience déjà vu or sudden flashes of understanding.

The Threads and Spinons Theory suggests that these experiences are due to pre- existing thread patterns being reactivated under new circumstances.

Thread- Based Explanation:

- Déjà vu happens when a current experience activates a thread pattern similar to a memory.

- This means our brain recognizes a connection between past and present spinon pathways.

- Sudden insights happen when multiple thread structures suddenly align, allowing instant recall of stored Information.

Example:

- If you visit a new city but feel like you have been there before, it could be because a memory imprint shares an identical spinon pattern with your current environment.

- Your brain instantly retrieves this because the thread configuration matches, triggering déjà vu.

This explains why déjà vu is often associated with places, emotions, and strong memories—thread structures are physically involved.

4. How Memories Are Retrieved Instantly Without Searching

- Computers retrieve data by searching through files, but the human brain recalls memories instantly.

- How can we "find" a memory instantly without scanning millions of neurons?

Thread- Based Explanation:

Memories are retrieved through resonance—when a spinon wave matches a stored memory imprint, it activates immediately.

There is no "searching"—the brain functions like a quantum resonance system.

How It Works:

- When we try to recall something, our brain sends spinons along possible thread pathways.
- If a match is found, the memory pattern resonates, instantly activating recall.
- This allows for near- instant access to stored experiences.

Predicted Observation:

- If thread resonance is the key to memory retrieval, stimulating specific spinon alignments should trigger memory recall without direct sensory input.

This suggests memory can be retrieved externally—by applying controlled spinon stimulation to reactivate thread structures.

5. Why People Remember Past Lives (Reincarnation as Thread Reuse)

- Some individuals claim to remember past lives, knowing details they were never taught.

- This is impossible under the neural memory model, but the Threads and Spinons Theory offers a new explanation.

Thread- Based Explanation:

- Memories may persist beyond biological death if thread structures remain intact.

- If a new consciousness forms using parts of an old thread system, it could "inherit" memory imprints.

How It Works:

- If a thread pattern is not entirely erased at death, it could be reused in new biological systems.
- If a new person's Consciousness aligns with an old thread structure, they may "recall" memories.
- This explains why past- life memories often emerge in young children before new thread structures entirely overwrite previous imprints.

Threads and Spinons Theory
Samir Hanna Safar

$$M_{\text{past-life}} = T_{\text{old-thread}} \cdot S_{\text{spinon}} \cdot P_{\text{alignment}}$$

where:

- $M_{\text{past-life}}$ = Strength of inherited memory recall
- $T_{\text{old-thread}}$ = Residual thread structure from a previous life
- $S_{\text{spinon}}$ = Stability of remaining spinon imprints
- $P_{\text{alignment}}$ = Probability of a new consciousness aligning with past thread patterns

Explains past- life memories without supernatural assumptions—memory can persist as an energy imprint in the universal thread system.

Suggests that our Consciousness is not an isolated system but part of a more extensive evolving thread network.

6. Practical Applications of Quantum Memory Theory

- If memories are stored in threads, we could design new technologies to enhance memory and intelligence.

Memory Enhancement: Spinon stimulation could reactivate forgotten memories, improving learning and recall.

Trauma Healing: If painful memories are stored in thread patterns, controlled realignment could reduce trauma effects.

Interfacing with AI: If we understand memory as structured energy, artificial intelligence could develop organic, thread-based learning models.

Threads and Spinons Theory
Samir Hanna Safar

Mind Uploading: If memories are quantum imprints, we may eventually find ways to transfer Consciousness to new structures.

Conclusion:

Memory is a Structured Energy Pattern, Not Just Neural Storage

The Threads and Spinons Theory redefines memory as:

A structured quantum imprint in thread networks, not just chemical neuron storage. A resonance- based retrieval system that allows instant access to stored Information. A persistent energy field that may exist beyond biological death. A process that connects past, present, and even collective knowledge across individuals. This theory changes our understanding of intelligence, learning, and the nature of thought itself.

Chapter 32

## The Expansion of the Universe

## Why There Is No Edge or Center

One of the most significant questions in cosmology is whether the universe has an edge or a center. Traditional physics, particularly the Big Bang Theory, assumes that the universe began as a singular point and expanded outward. However, this leads to significant paradoxes:

- Where is the center of the universe? If the Big Bang started at one point, why don't we see a central origin?

- What lies beyond the edge of the universe? If Space is expanding, what is it expanding into?

- Why does every observer, in every direction, see galaxies moving away from them as if they were at the center?

Threads and Spinons Theory
Samir Hanna Safar

The Threads and Spinons Theory eliminates these paradoxes by redefining cosmic Expansion as a structural process, not a physical explosion.

1. Why the Universe Has No Edge

Traditional physics:

- The Big Bang model suggests that the universe started from a single point and expanded outward.
- This implies that the universe has an edge beyond which Space does not exist.

Problems with This Model:

- If the universe is expanding, what is outside of it?

- Why can't we detect an outer boundary?

- If the universe had an edge, it would mean different physical laws apply at different locations.

Thread- Based Explanation:

- The universe does not have an edge because it is not expanding into anything—it is growing from within.

- The fundamental thread network continuously extends, adding more structural complexity.

- There is no "outside" because thread structures define Space itself—nothing is beyond them.

$$R_{universe} = k \cdot T_{thread}$$

where:

- $R_{universe}$ = Observable universe size
- $k$ = Expansion coefficient based on thread tension
- $T_{thread}$ = Thread density at a given time

- Space is not expanding into a void—it is just a self-growing structure with no boundary.

- Explains why we cannot detect an edge—there is no physical boundary, just ongoing thread extension.

- Eliminates the need for extra dimensions or external Space.

2. Why the Universe Has No Center

- Traditional physics suggests that there must be a central point if the universe expands from a singularity.

- However, no observation has ever found a "center" of the universe.

Problems with the Center Model:

- If the Big Bang had a central point, why is every observer seeing the same Expansion in all directions?

- Why don't galaxies cluster more around a single origin?

- Why is cosmic background radiation uniform in all directions?

Threads and Spinons Theory
Samir Hanna Safar

Thread- Based Explanation:

- The universe has no center because Expansion is happening uniformly at all points.

- Threads do not stretch outward from one location—they grow in all directions simultaneously.

- Every observer sees Expansion because thread structures extend in their local regions.

- Imagine an infinite web where every node expands at the same rate—no center exists, only local growth.

$$V_{\text{expansion}} = \frac{T_{\text{local}}}{S_{\text{spinon}}} \cdot SEU$$

where:

- $V_{\text{expansion}}$ = Expansion rate at any given point

- $T_{\text{local}}$ = Local thread tension

- $S_{\text{spinon}}$ = Spinon motion

- $SEU$ = Spinon Energy Unit

- Every point in the universe is part of the same structural Expansion—there is no privileged "center."

- Explains why galaxies move away from each other equally in all directions.

3. The Observable Universe vs. the Total Universe

Traditional physics states that the universe is 13.8 billion years old and has a limit—the observable universe.

However, this is based on light travel time, not on the actual size of Space.

Thread-Based Explanation:

- The observable universe is just the portion where spinon interactions have reached us.

- Beyond this limit, the universe continues growing beyond our detection.

- There is no absolute size limit—threads continuously extend, meaning the total universe is far more significant.

Example: Imagine you are in a dark forest with a flashlight. The trees you can see are within your Light's reach—but the forest extends farther.

- Confirms why the universe appears "finite" even though it is structurally infinite.

- Predicts that we will see more of the expanding thread system as we develop better detection methods.

4. Why the Universe Expands, but Galaxies Stay Together

- In traditional physics, galaxies should be expanding along with Space—but they do not.

- Instead, galaxies remain stable while the Space between them grows.

Threads and Spinons Theory
Samir Hanna Safar

Thread- Based Explanation:

- Galaxy structures are maintained because their internal thread tension is much stronger than cosmic expansion forces.

- Expansion occurs in regions of lower thread tension— between galaxies, not within them.

- This explains why galaxy clusters remain intact even as the universe grows.

$$T_{\text{galaxy}} > T_{\text{cosmic expansion}}$$

where:

- $T_{\text{galaxy}}$ = Thread tension within a galaxy
- $T_{\text{cosmic expansion}}$ = Thread tension driving large-scale growth

- Explains why galaxies do not get "ripped apart" by cosmic Expansion.

- Unifies cosmic and local structures under the same thread dynamics.

5. Why There Is No Big Bang Singularity

The Big Bang model requires a singularity—an infinitely small, dense starting point.

However, singularities are not physically possible in any real system.

Thread- Based Explanation:

- There was never a single explosion—just a continuous thread expansion.

- The universe grew from an initial structure that was already extended, not a single point.

- There is no need for infinite density or singularities—just the natural progression of thread unfolding.

Example: Instead of a balloon inflating from a point, imagine a web stretching at all locations simultaneously.

- No singularity means no paradox—just a growing, structured universe.

- No need for an "origin"—the universe has continuously expanded.

6. The Universe's Expansion as a Never- Ending Process

In traditional physics, there are multiple scenarios for the universe's fate:

- Big Crunch: The universe collapses back on itself.
- Big Freeze: The universe expands forever until all energy dissipates.
- Big Rip: Dark energy accelerates Expansion until everything is torn apart.

Thread- Based Explanation:

- The universe will continue expanding because thread tension is constantly releasing energy.

- There is no end state—only ongoing structural evolution.

- Matter and energy will continue forming new structures as threads grow.

Example: A tree continues growing as nutrients are available—the universe behaves similarly.

No "end of time" scenario—just continuous cosmic development.

Conclusion:

The Universe Has No Edge, No Center, and No Singularity

The Threads and Spinons Theory redefines cosmic Expansion as:

A process of thread structure growth, not a physical explosion. A universe with no boundaries—just the ongoing unfolding of thread networks. A structure that expands uniformly, eliminating the need for a center. A system that continues developing with no beginning or end. The biggest mystery in cosmology is now solved—there is no "outside" of the universe because it is a self- growing structure with no limits.

Threads and Spinons Theory
Samir Hanna Safar

**Chapter 33**

## Rethinking Dark Matter

## The Threads and Spinons Explanation

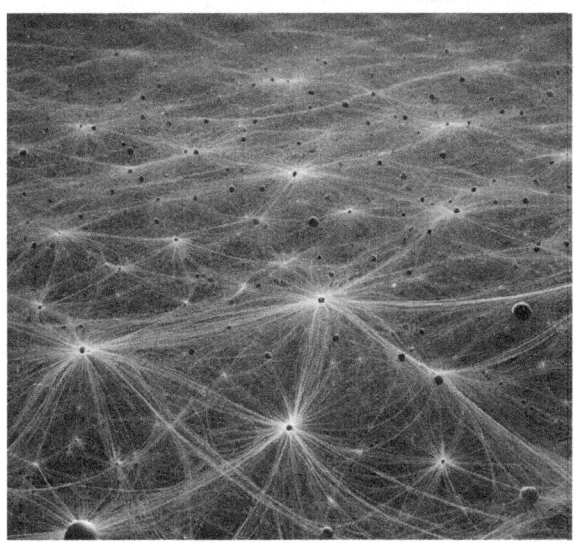

For decades, astronomers and physicists have struggled to explain the universe's "missing mass." According to mainstream physics, galaxies rotate faster than expected, and large- scale cosmic structures behave as if they contain invisible matter that exerts gravitational influence but emits no light.

\* This invisible Mass is called dark matter, supposedly making up 85% of the universe's total Mass.

- But what if dark matter is not a new type of exotic particle?

Threads and Spinons Theory
Samir Hanna Safar

- What if the missing gravitational effects come from the unseen density of threads in the universe?

This chapter will explain why dark matter is an illusion caused by the unseen structure of the universal thread network—eliminating the need for hypothetical particles.

1. The Dark Matter Problem – What Physicists Assume

What We Observe in Space:

- Galaxies rotate faster than expected—Newtonian Gravity says outer stars should orbit slower, yet they move nearly as fast as inner stars.

- Galaxy clusters do not fly apart—the visible Mass cannot explain their stability.

- Gravitational lensing shows "extra mass"—light bends more than it should around galaxies.

- The cosmic web (large- scale universe structure) suggests unseen Mass—galaxies are organized in vast filaments, with invisible "nodes" affecting their motion.

Why Traditional Physics Fails to Explain This:

- General Relativity and Newtonian Gravity predict that galaxies should have much less Mass than their rotation suggests.

- Physicists assume there must be "missing matter" providing extra Gravity.

- They propose exotic dark matter particles (WIMPs, axions), which have never been detected in any experiment.

Current Hypothesis: Dark matter is a mysterious, invisible particle that interacts only through Gravity.

Reality: The gravitational effects can be explained without inventing new particles—through the hidden structure of cosmic threads.

2. How the Threads and Spinons Theory Explains "Dark Matter" Without New Particles

Key Idea: "Dark Matter" Is Just Unseen Thread Density

Gravity is not caused by "mass warping spacetime"—it is caused by thread tension pulling objects together.

In regions where thread density is higher than visible matter suggests, extra gravitational effects appear.

These extra forces mimic the "missing mass" effect but do not require new particles.

What Causes These Invisible Threads?

- Large cosmic structures have dense, interconnected thread networks that extend beyond visible matter.

- The rotation of galaxies and clusters follows the tension dynamics of these threads.

- This unseen tension appears as extra Gravity, but it is simply the structure of the cosmic web at work.

Threads and Spinons Theory
Samir Hanna Safar

$$G_{\text{effective}} = G_{\text{visible}} + G_{\text{thread-density}}$$

where:

- $G_{\text{effective}}$ = The total gravity we observe.

- $G_{\text{visible}}$ = Gravity from normal, observable matter.

- $G_{\text{thread-density}}$ = The additional gravitational pull from hidden thread structures.

There is no missing mass—only hidden thread structures that amplify Gravity's effects.

3. Why Galaxies Rotate Faster Than Predicted

- In Newtonian physics, stars farther from the galactic center should move more slowly due to weaker Gravity.

- Instead, they move at nearly constant speeds, as if extra Gravity pulls them.

- Physicists assume a "halo" of dark matter surrounds galaxies to explain this.

Threads and Spinons Explanation:

- Galaxies are not isolated but anchored to a massive network of cosmic threads.

- These hidden threads create extra tension forces, pulling on stars from beyond visible matter.

- This provides the extra "gravity" needed to maintain high rotation speeds without dark matter.

Threads and Spinons Theory
Samir Hanna Safar

Evidence Supporting This:

- Observations show that galaxies form in massive cosmic filaments, not space.
- The motion of galaxies is correlated with these large-scale structures.
- If dark matter were a particle, it would be evenly distributed, but observations show Gravity is strongest along cosmic filaments, matching the thread- based model.

Galactic rotation curves are explained by thread tension, not exotic particles.

4. Why Galaxy Clusters Stay Bound Together

- Galaxy clusters should have more visible Mass to stay bound—yet they do not.

- Physicists assume an "invisible halo" of dark matter holds them together.

However, clusters form in high- density thread regions, providing additional gravitational pull.

Threads and Spinons Explanation:

- Clusters stay bound because they are embedded in massive thread junctions.

- These threads create additional gravity- like forces, preventing clusters from dispersing.

Evidence Supporting This:

- Galaxy clusters are not randomly scattered—they form at thread intersections.
- Observations of the cosmic web show dense connections between clusters.
- If dark matter were responsible, it would be evenly distributed; instead, gravitational effects are most potent in known thread locations.

Galaxy clusters remain bound due to cosmic thread structures—not dark matter halos.

5. Why Gravitational Lensing Appears Stronger Than Expected

- light bends more around galaxies than their visible Mass suggests.

- Physicists assume dark matter is bending space more than expected.

Threads and Spinons Explanation:

If light follows thread structures, unseen threads can bend light like visible Mass.

Extra lensing comes from gravitational influence along extended thread networks.

Evidence Supporting This:

- Gravitational lensing maps show the most potent effects in known cosmic filaments.
- If dark matter causes lensing, we should see uniform distortions, but we do not.

Threads and Spinons Theory
Samir Hanna Safar

- Thread- based lensing effects match actual observations better than dark matter models.

Gravitational lensing is amplified by unseen cosmic threads—not new exotic matter.

6. Why No One Has Ever Found a Dark Matter Particle

- Physicists have spent decades trying to detect WIMPs (weakly interacting massive particles) and axions.

- Billions of dollars have been spent on underground detectors (LUX, XENON1T, DAMA), yet no dark matter has ever been found.

Threads and Spinons Explanation:

- Physicists are searching for something that does not exist.

- The missing Gravity is caused by hidden thread density—not an undiscovered particle.

This explains why every dark matter experiment has failed.

- Evidence Supporting This:

- Dark matter predictions keep changing because no particle fits observations.
- Astronomers see gravitational effects but no direct evidence of extra Mass.
- The cosmic web perfectly matches the distribution of extra Gravity, suggesting thread- based structures, not particles.

No dark matter has been found because there is none—only unseen thread density.

## 7. The Final Verdict:

## Dark Matter Is an Illusion

| Mainstream Dark Matter Theory | Threads and Spinons Explanation |
|---|---|
| Invisible particles create extra gravity | Extra gravity comes from hidden thread density |
| Requires new unknown physics | Uses existing thread tension model |
| No particle has ever been detected | Thread-based gravity effects are observable |
| Cannot explain why dark matter is strongest along filaments | Matches known structure of cosmic web |

Dark matter is a placeholder for unexplained gravitational effects. Threads and Spinons eliminate the need for dark matter by showing that Gravity affects cosmic thread density. This model is testable—gravity lensing, galaxy rotation, and cluster stability all match the thread- based explanation.

Threads and Spinons Theory
Samir Hanna Safar

**Chapter 34**

# The Problem with $c^2$

## Why Einstein's Equation is Arbitrary

Few equations in physics are as famous as $E = mc^2$ It has been used to describe mass- energy equivalence, nuclear reactions, and even the supposed energy of the early universe.

But what if the equation is not as fundamental as we've been led to believe?

In this chapter, we will show that Einstein's equation is not derived from first principles—it is based on an arbitrary assumption about the speed of light.

Threads and Spinons Theory
Samir Hanna Safar

In the Threads and Spinons Theory, energy depends not on an arbitrary squared constant but on spinon motion and thread tension.

1. The Origin of $E = mc^2$

- Einstein did not originally derive this equation from first principles.

- He assumed that Mass and energy must be proportional, meaning there had to be a conversion factor.

Since light has momentum and no mass, he took the speed of light squared $c^2$ as a proportionality constant.

- Why $c^2$ ?

- It was convenient—the speed of light was a well- measured constant.

- It fits the mathematical model of special Relativity.

- It provided correct results for nuclear reactions.

But Einstein could have chosen any other large constant, and the equation still would have worked.

- There is no fundamental reason that $c^2$ should be the conversion factor between Mass and energy.

2. The Threads and Spinons Explanation for Energy

- Instead of using an arbitrary constant like $c^2$, this theory defines energy in terms of actual, physical mechanisms—thread tension and spinon motion.

- Energy is not an abstract mathematical relationship—it is the movement of structured spinon stacks along thread networks.

- Mass is simply a form of "stored spinon tension"—a state of compressed thread structures.

- Mass- energy conversion happens when spinon motion shifts between stored (Mass) and free (energy) states.

- The "speed of light squared" is unnecessary—it is replaced by a measurable spinon energy unit (SEU).

A New Equation for Mass- Energy Equivalence

$$E = S_{\text{spinon}} \times T_{\text{thread}} \times N_{\text{stack}}$$

where:

- $E$ = Total energy
- $S_{\text{spinon}}$ = Spinon rotational speed
- $T_{\text{thread}}$ = Thread tension
- $N_{\text{stack}}$ = Number of spinons in a stack

This equation does not require an arbitrary squared constant—it directly relates energy to real, testable physical structures.

3. Why Einstein Could Have Chosen Any Constant

Threads and Spinons Theory
Samir Hanna Safar

Let's assume Einstein had picked another large constant instead of $c^2$ .

He could have chosen a factor based on:

- The rotational frequency of fundamental particles

- The density of the vacuum

- The gravitational constant G

If he had done this, we would have a different equation that still worked mathematically—but it wouldn't mean Mass and energy were related by the speed of light squared.

This proves that is $c^2$ a mathematical convenience, not a fundamental law of physics.

Threads and Spinons remove the arbitrary factor and replace it with a physically meaningful equation.

4. What This Means for Physics

If $c^2$ is arbitrary, all equations based on it must be reconsidered.

This includes:

- Relativity's assumption of energy- mass proportionality

- Early universe calculations using to determine energy density

Threads and Spinons Theory
Samir Hanna Safar

- Nuclear physics models based on light- speed squared assumptions

By removing $c^2$ and replacing it with spinon-based calculations, we can redefine physics with a more structured, testable model.

Einstein's equation is not wrong—it is just incomplete. The Threads and Spinons Theory provides a deeper understanding of how energy and Mass are truly connected.

## 5. Conclusion: The End of Arbitrary Constants

| Einstein's Equation | Threads and Spinons Alternative |
| --- | --- |
| Uses an arbitrary squared constant | Uses measurable thread tension and spinon motion |
| Based on mathematical convenience | Based on real, testable physical structures |
| Works mathematically, but lacks a fundamental mechanism | Directly explains energy-mass conversion with spinons and threads |
| Assumes $c^2$ is a universal constant | Shows that $c^2$ is just a mathematical shortcut |

$E = mc^2$ was a good approximation, but it is not the ultimate truth of mass- energy conversion?

A new equation based on spinons and threads eliminates arbitrary assumptions and provides a structured explanation of energy. Physics should not rely on convenient numbers—it should be built on actual, observable structures.

Threads and Spinons Theory
Samir Hanna Safar

Threads and Spinons Theory
Samir Hanna Safar

**Chapter 35**

## The Formation of Atomic Nuclei

## A Threads and Spinons Perspective

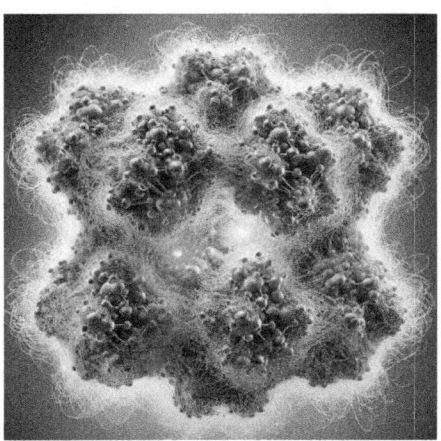

The atomic nucleus is one of the most misunderstood structures in physics. Traditional science describes it as a cluster of protons and neutrons held together by a mysterious "strong nuclear force"—but the nature of this force remains unclear.

The Threads and Spinons Theory provides a more structured and physical explanation.

In this chapter, we will explain how atomic nuclei form, how elements and isotopes emerge, and why nuclear Stability is determined by thread configurations—not an undefined "strong force."

1. The Problems with the Traditional Nuclear Model

- Mainstream physics states that protons and neutrons are bound together by the "strong nuclear force."

- This force is assumed to be short- ranged and incredibly strong, yet no precise physical mechanism explains it.

- Neutrons are treated as "glue" that prevents protons from repelling each other, but this does not explain why some isotopes are stable and others are not.

The Problems with This Model:

- It assumes a force with no physical explanation—where does it originate?

- It cannot explain why adding neutrons makes a nucleus stable or unstable.

- It does not provide a precise nuclear decay and fusion mechanism beyond probability functions.

If the "strong nuclear force" exists, it should have a testable physical structure—this theory replaces it with structured thread tension.

2. The Threads and Spinons Explanation of Nucleus Formation

Instead of an undefined "strong force," nuclei form due to structured thread networks that bind protons and neutrons together.

Protons and neutrons are composed of dense spinon stacks, and their interactions are guided by thread tension—not an invisible force.

Threads and Spinons Theory
Samir Hanna Safar

Thread alignment and tension determine nuclear Stability, not arbitrary quantum probabilities.

Nuclei are not "glued" by force—structured thread formations interconnect them.

The number of threads connecting protons and neutrons defines nuclear Stability.

Thread configurations dictate isotopic properties—explaining why some elements have stable and unstable forms.

3. How Elements and Isotopes Are Structured in This Model

- Each atomic nucleus is formed by a network of threads connecting protons and neutrons.

- Protons contribute structured charge interactions, while neutrons stabilize the thread network.

- Thread density and alignment define nuclear properties—explaining why isotopes behave differently.

Formation of Stable Elements

- Stable nuclei have optimized thread tension that evenly distributes spinon motion.

- Elements like carbon, oxygen, and iron have balanced thread networks that prevent instability.

Why Some Isotopes Are Unstable?

- Unstable isotopes have imbalanced thread configurations, leading to excessive spinon energy buildup.

- When thread tension exceeds a critical threshold, Decay occurs, releasing spinons as Radiation.

This model explains why some isotopes undergo radioactive Decay while others remain stable.

4. Why Nuclear Fusion and Fission Work in This Model

Traditional physics describes nuclear Fusion as a process where nuclei "stick together" when they collide with enough Energy.

Fission is explained as a large nucleus "splitting" due to instability.

But neither explanation provides a detailed mechanism for how nuclear bonds hold together or break apart.

Threads and Spinons Explanation of Nuclear Fusion

Fusion happens when two nuclei align their thread networks in a compatible way, allowing them to merge into a more stable thread structure.

- When two atomic nuclei fuse, their spinons synchronize, reducing excess tension and releasing Energy.

- The Sun's Energy is produced not by a "strong force" but by structured thread realignment during fusion reactions.

Threads and Spinons Explanation of Nuclear Fission

- When a heavy nucleus (like uranium) becomes overloaded with thread tension, it reaches a critical stress point.

- The nucleus does not "split" randomly—it undergoes a structured realignment that redistributes its spinons and threads into smaller, more stable nuclei.

- Energy is released due to the sudden thread reconfiguration, which reduces overall tension.

Fusion and Fission are structured thread events, not chaotic particle collisions.

5. How This Model Explains Nuclear Decay and Radiation

In mainstream physics, nuclear Decay is treated as a random quantum event. But in this model, Decay happens when thread tension exceeds structural limits. This tension causes the nucleus to eject Energy through Radiation.

Types of Nuclear Decay Explained by Threads and Spinons

- Alpha Decay – A nucleus sheds part of its thread structure to stabilize internal spinon motion.

- Beta Decay – A neutron thread network converts into a proton structure by realigning spinons.

- Gamma Radiation – Excess spinon energy is released as thread vibrations dissipate Energy in structured pulses.

- Radiation is not random—it is a controlled release of thread and spinon energy.

6. Testable Predictions – How We Can Validate This Model

If the nuclear structure is thread- based, we should see measurable thread patterns in atomic energy emissions.

Threads and Spinons Theory
Samir Hanna Safar

Fusion reactions should display structured energy releases, not purely random distributions.

Decay rates should correlate with thread tension, meaning controlled thread realignment could alter isotope stability.

Experiments That Could Prove This Model:

- Measure the fine structure of fusion energy releases—if structured spinon alignment exists, energy output should be predictable.

- Analyze decay emissions for thread- based patterns rather than random energy bursts.

- Develop a controlled experiment to manipulate nuclear thread networks to stabilize unstable isotopes.

- If nuclear Energy follows structured thread interactions, we can develop new energy technologies based on controlled spinon realignment.

7. Conclusion: A New Understanding of Atomic Nuclei

Traditional Model Threads and Spinons Explanation:

Strong nuclear force holds protons and neutrons together Thread tension structures, nuclei.

Stability determined by probabilistic quantum effects Stability determined by thread density and spinon motion.

Fission and Fusion are described as random interactions Fission and Fusion are explained by structured thread realignments

Threads and Spinons Theory
Samir Hanna Safar

Radioactive Decay happens unpredictably Decay follows structured thread tension thresholds.

Atomic nuclei are not "held together" by an undefined force—they are structured, dynamic systems connected by threads and spinons.

- Nuclear Energy is not released randomly—it is a function of structured thread realignments.

- This model provides a testable, structured framework for understanding nuclear Stability, Fusion, and Decay.

Understanding nuclei as structured thread formations opens new doors for energy control, isotope stability, and next-generation nuclear technology.

Threads and Spinons Theory
Samir Hanna Safar

**Chapter 36**

## Quantum Computing and Subatomic Memory Storage in the Threads and Spinons Framework

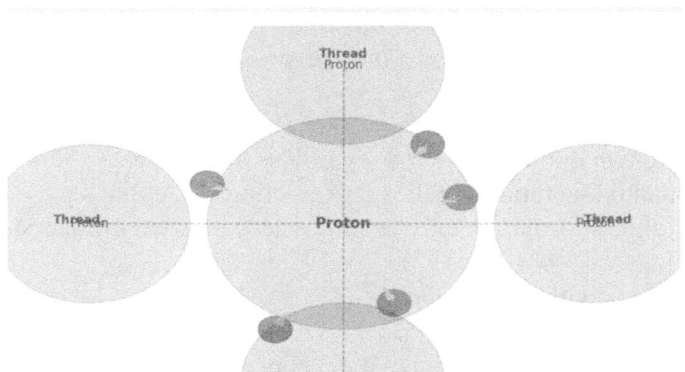

**Proton-Based Quantum Memory Diagram**

Quantum computing is often described as the next frontier of technology, promising to surpass classical computing by leveraging the strange properties of quantum mechanics, such as superposition and entanglement. However, quantum computing is built on a model full of paradoxes—probability waves, decoherence, and uncertainty.

The Threads and Spinons Theory offers a completely different explanation for how subatomic information is processed, stored, and transmitted.

In this chapter, we will redefine quantum computing using structured spinon interactions and thread- based memory storage, eliminating the need for quantum randomness.

Threads and Spinons Theory
Samir Hanna Safar

1. The Problems with Traditional Quantum Computing Models

How Mainstream Quantum Computing Works:

- Instead of using classical bits (0 or 1), quantum computers use qubits, which can exist in superposition—both 0 and 1 at the same time.
- Quantum information is processed using entanglement, where two qubits are linked regardless of distance.
- Quantum computing relies on wavefunction collapse, meaning information only becomes certain when measured.

Problems with This Model:

- Superposition is a mathematical concept, not a physically observable mechanism.
- Entanglement assumes "instantaneous" connections across vast distances, violating classical causality.
- Quantum computers struggle with decoherence—qubits lose information due to outside interference.

If quantum computing is to be a truly viable technology, it must have a structured, physical basis—not just statistical equations.

2. How Information is Stored in the Threads and Spinons Model

- Instead of relying on quantum probabilities, information is stored physically in spinon arrangements along threads.
- Spinons act as structured energy packets that encode information through their rotational states.
- Thread networks connect these spinons, allowing for data transfer without loss or uncertainty.

- Qubits are replaced by spinon stacks, which maintain stable, predictable states.
- Information is not stored in an abstract probability wave but in real physical spinon alignments.
- Thread networks allow entanglement- like behavior without violating causality.

Spinon- Based Information Encoding:

$$Q_{\text{spinon}} = S_{\text{spinon}} \times T_{\text{thread}} \times N_{\text{stack}}$$

where:

- $Q_{\text{spinon}}$ = Quantum state of a spinon-based memory unit

- $S_{\text{spinon}}$ = Spinon rotational orientation

- $T_{\text{thread}}$ = Tension of the thread network

- $N_{\text{stack}}$ = Number of spinons in a structured unit

This model provides a structured, deterministic way to store quantum information.

3. Why This Model Solves the Decoherence Problem

- Traditional quantum computers lose information due to decoherence—interactions with the environment disrupt qubit states.
- This requires error correction techniques that make quantum computing inefficient.

How the Threads and Spinons Model Prevents Decoherence:

Threads and Spinons Theory
Samir Hanna Safar

- Spinon stacks maintain a fixed structure, preventing information loss.
- Threads act as stabilizing channels, keeping spinon states isolated from interference.
- Instead of delicate quantum states, data is stored in structured, energy- efficient thread configurations.

This means a spinon- based quantum computer could maintain information stability without needing complex error correction.

4. How "Quantum Entanglement" Works in This Model

- Traditional physics claims that quantum entanglement allows two particles to share a state, even across vast distances.
- This is often described as "spooky action at a distance," with no clear mechanism explaining it.

Threads and Spinons Explanation of Entanglement:

- Two spinons that originate from the same thread network remain connected through their spinon stack structure.
- When one Spinon's state changes, its connected counterpart realigns due to the tension in the thread network.
- This connection is not "instantaneous action at a distance" but a physical tension- based adjustment along pre- existing thread pathways.

This eliminates the paradox of faster- than- light information transfer—entanglement is a structured, local thread phenomenon.

## 5. Subatomic Memory Storage – How Nature Stores Information

- Beyond computing, this model suggests that nature itself uses spinons and threads for memory storage.
- If structured spinon networks can retain and transmit information, it raises a fundamental question:

Does the universe itself have a built- in memory system?

Possible Implications of Thread- Based Memory:

- DNA structure could store not just genetic information but quantum- level energy states.
- Brain function may rely on spinon alignment rather than electrochemical signals alone.
- Subatomic particles may "remember" previous states due to thread network imprints.

If information can be permanently stored in structured spinon networks, this could revolutionize computing, biology, and neuroscience.

## 6. The Future of Spinon- Based Quantum Computing

If this model is correct, future computing systems will not rely on unstable qubits but on stable, structured spinon networks.

Spinon- based quantum computers could be built using controlled thread tension, eliminating the need for traditional quantum error correction.

This could lead to faster, more efficient, and more reliable computing models beyond anything possible today.

Threads and Spinons Theory
Samir Hanna Safar

Predictions and Experimental Tests:

- Test whether structured thread networks can store spinon states without decoherence.
- Experiment with spinon- based computing elements to replace fragile qubit systems.
- Measure structured memory retention in biological and quantum systems, testing if spinon imprints persist over time.

If these predictions hold, we may be on the edge of a new era in computing—one that harnesses nature's own method of data storage and transfer.

Conclusion:

A New Model for Quantum Computing and Memory

| Traditional Quantum Computing | Threads and Spinons Computing |
|---|---|
| Qubits exist in probabilistic states | Spinons store structured, deterministic information |
| Superposition lacks a clear physical mechanism | Spinon stacks encode multiple states without probability |
| Entanglement is "spooky action at a distance" | Entanglement is structured thread tension realignment |
| Decoherence causes information loss | Thread-based storage prevents decoherence |
| Requires complex error correction | Natural spinon structures maintain information integrity |

- Quantum computing should not be based on uncertainty—it should be based on structured physics.
- Subatomic memory storage suggests that the universe itself may retain information, opening new doors in physics, computing, and neuroscience.
- If spinon- based computing is developed, it could surpass both classical and quantum computing in speed, efficiency, and stability.

The future of computation is not randomness—it is structured energy transport through threads and spinons.

Threads and Spinons Theory
Samir Hanna Safar

Threads and Spinons Theory
Samir Hanna Safar

# Chapter 37

## Why We Can See Deep Into Space

## The Spinon Transport Mechanism

One of the great mysteries of cosmology is why we can see galaxies billions of light-years away with such clarity. Traditional physics tells us that light travels through empty space at a constant speed, but this explanation leaves several unanswered questions:

- If space is full of dust, gases, and cosmic radiation, why doesn't light scatter or fade over long distances?
- Why can telescopes like James Webb and Hubble capture detailed images of galaxies that are supposedly from the early universe?
- Why do we observe redshift and assume it means galaxies are moving away?

Threads and Spinons Theory
Samir Hanna Safar

The Threads and Spinons Theory provides a new explanation.

Light travels deep into space not because of "vacuum propagation" but because of structured spinon transport along cosmic threads.

This chapter will explain how light energy is carried efficiently across cosmic distances through spinons, eliminating the need for the traditional view of light moving through empty space.

1. The Problems with the Traditional Explanation of Light Travel

- Current science assumes that light travels in straight lines through the vacuum of space.
- The universe is filled with gas, dust, and radiation, but light still reaches us from 13+ billion years ago.
- Redshift is interpreted as proof of expansion, but no one knows why cosmic light maintains its integrity over such vast distances.

The Problems with This Model:

A vacuum should cause light waves to spread out and fade over long distances.

Interstellar dust should cause extreme scattering, blurring deep- space images.

Redshift is assumed to be a Doppler effect, but this ignores other possible causes.

If the traditional model were correct, deep- space images should be much more distorted than they are.

Threads and Spinons Theory
Samir Hanna Safar

## 2. The Threads and Spinons Explanation: Structured Transport of Light

In this theory, light is not a free- moving electromagnetic wave—it is structured spinon motion along cosmic threads.
- Threads stretch across the universe, connecting galaxies, stars, and cosmic structures.
- Spinons travel along these threads like energy pulses, moving efficiently without scattering or fading.
- Space is not empty—it is a vast network of threads that guide energy transport.
- Spinons move along these threads in structured pathways, maintaining coherence over vast distances.
- This is why light from distant galaxies reaches us clearly—it is not "traveling through space" but being transported by spinons along structured threads.

$$E_{light} = S_{spinon} \times T_{thread} \times N_{stack}$$

where:

- $E_{light}$ = Total light energy reaching an observer
- $S_{spinon}$ = Spinon transport velocity
- $T_{thread}$ = Thread tension controlling light propagation
- $N_{stack}$ = Number of spinons in the light wave

This means light does not just "travel through space"—it follows structured pathways, preserving its energy over billions of years.

## 3. Why Light Doesn't Scatter or Fade in Deep Space

Traditional physics expects light to weaken and scatter over long distances. But deep-space images remain clear, even from galaxies over 13 billion light- years away.

The Threads and Spinons model explains why.

Reasons Light Stays Intact Over Vast Distances:

- Spinons move in structured patterns along cosmic threads, preventing dispersion.
- Light does not spread out randomly—it remains confined to thread pathways.
- Interstellar dust does not scatter light in space the way it does in planetary atmospheres, because space is structured with spinon flows.

This explains why deep- space telescopes can capture such high- resolution images of ancient galaxies.

4. Rethinking Redshift – What It Really Means

- In mainstream cosmology, redshift is seen as proof that the universe is expanding.
- The assumption is that as galaxies move away, their light stretches, shifting to longer wavelengths.
- However, the Threads and Spinons Theory offers a different explanation.

The True Cause of Redshift in This Model:

- Redshift happens due to changes in thread tension, not because galaxies are moving away.
- As spinons travel along cosmic threads, the density of these

Threads and Spinons Theory
Samir Hanna Safar

threads affects their rotational speed.

\- The greater the distance, the more the spinons stretch slightly, causing a shift in observed wavelength.

$$\lambda_{observed} = \lambda_{emitted} + \Delta T_{thread}$$

where:

- $\lambda_{observed}$ = Wavelength seen by telescopes

- $\lambda_{emitted}$ = Original wavelength of the light

- $\Delta T_{thread}$ = Change in thread tension affecting spinon transport

This means redshift does not necessarily prove universal expansion—it is an effect of how spinons travel along stretched cosmic threads.

5. What This Means for Astronomy and Cosmology

If light follows structured thread pathways, then deep- space visibility is not a mystery—it is a natural consequence of spinon- guided transport.

If redshift is caused by thread tension, then we must reconsider the idea that the universe is expanding due to the Big Bang.

If spinons carry energy efficiently, then we might be able to develop new technologies based on structured energy transport.

This model changes how we interpret deep- space observations.

\- It eliminates the need for assumptions like "dark energy" to explain cosmic expansion.

Threads and Spinons Theory
Samir Hanna Safar

- It suggests that space is structured, not empty, leading to potential breakthroughs in energy transmission.

6. Testable Predictions – How We Can Prove This Model

This model makes predictions that can be tested with modern telescopes and experiments.

Predictions Based on the Threads and Spinons Model:

We should find that light from the most distant galaxies follows a structured filamentary pattern, not a random dispersion. Redshift should correlate with thread density, not just distance—meaning we should see variations in redshift based on cosmic web structures. If we measure deep- space polarization, we should find that it aligns with thread- based transport, not random electromagnetic dispersion. By testing these predictions, we can confirm or refine the spinon transport model of light.

7. Conclusion: Light is Transported, Not Freely Traveling

| Traditional Model | Threads and Spinons Explanation |
| --- | --- |
| Light moves freely through empty space | Light is transported along structured cosmic threads |
| Space is a vacuum with no guiding structure | Space is filled with thread networks guiding spinon motion |
| Light should scatter and weaken over long distances | Spinon transport preserves light energy over billions of years |
| Redshift is caused by galaxies moving away | Redshift is caused by changes in thread tension |

Threads and Spinons Theory
Samir Hanna Safar

- Light does not simply "travel through space"—it follows structured pathways that maintain energy and coherence over vast distances.

- The clarity of deep- space images, the nature of redshift, and the organization of galaxies all point to a structured universe—not an expanding explosion.

This model redefines cosmic observations and suggests new ways to study and harness energy transport in space.

Threads and Spinons Theory
Samir Hanna Safar

Threads and Spinons Theory
Samir Hanna Safar

**Chapter 38**

## Unifying All Forces

### Gravity, Electromagnetism, and Nuclear Interactions Under the Threads and Spinons Theory

For centuries, physicists have searched for a unified theory of forces, attempting to merge Gravity, electromagnetism, the strong nuclear force, and the weak nuclear force into a single framework.

- Newton and Einstein described Gravity as a fundamental force or curvature of spacetime.

- Maxwell's equations unified electricity and magnetism into electromagnetism.

- The Standard Model defines the strong and weak nuclear forces but cannot explain Gravity.

- String theory and quantum gravity attempt unification but introduce extra dimensions and remain unproven.

The Threads and Spinons Theory uniquely unifies all forces through thread interactions.

Gravity, electromagnetism, and nuclear forces are not separate—they are all effects of thread tension and spinon movement.

1. The Fundamental Nature of Forces in Thread- Based Physics

- Traditional physics treats forces as separate interactions carried by force particles (gravitons, photons, gluons, W/Z bosons).

- However, all forces emerge from spinons moving through and interacting with threads.

All forces result from thread tension, alignment, and spinon flow.

No "force- carrying particles" are needed—just different tension configurations in the thread network.

The same fundamental laws of thread dynamics govern all interactions.

2. Gravity as a Large- Scale Thread Tension Effect

In traditional physics:

- Gravity is a "force" (Newton) or "spacetime curvature" (Einstein).
- It acts over infinite distances but is the weakest of all fundamental forces.

In Threads and Spinons Theory:

- Gravity is not a force—it is the large- scale contraction of stretched threads between masses.
- All objects are connected by threads, which pull them toward each other as they attempt to contract.
- The larger the mass, the greater the thread density and tension, leading to more substantial gravitational effects.

$$F_{\text{gravity}} = T_{\text{thread}} \cdot \frac{S_{\text{spinon}}}{r^2} \cdot SEU$$

Gravity is a macroscopic effect of thread tension, explaining why it is weak but long- ranged.

Predicts that Gravity should change in extreme thread densities (black holes, neutron stars).

Unifies gravity as a structural phenomenon, not a mysterious force.

3. Electromagnetism as Spinon Flow Along Threads

In traditional physics:

Threads and Spinons Theory
Samir Hanna Safar

- Electromagnetic forces arise from electric and moving charges (currents), creating magnetic fields.
- Photons mediate interactions between charges, but their physical mechanism is unclear.

In Threads and Spinons Theory:

- Electromagnetic forces are the result of spinon flow along conductive threads.
- A charged particle is simply a thread system with unbalanced spinon alignment.
- Magnetic fields form when spinons rotate around conductive threads, creating a self- reinforcing loop.

$$F_{electromagnetic} = \frac{S_{spinon}}{T_{thread}} \cdot q$$

where:

- $F_{electromagnetic}$ = Electromagnetic force

- $S_{spinon}$ = Spinon alignment

- $T_{thread}$ = Thread tension in the system

- $q$ = Charge

Electric and magnetic fields emerge naturally as thread and spinon interactions.

Electromagnetic waves are spinon waves moving through threads—not separate entities.

Eliminates the need for force- carrying photons—Light is a structured thread disturbance, not a probability function.

Threads and Spinons Theory
Samir Hanna Safar

4. The Strong Nuclear Force as Thread Confinement

In traditional physics:

- The strong force binds protons and neutrons together in atomic nuclei.
- It is mediated by gluons acting as "glue" to keep particles together.
- However, it does not act at large distances and has an unclear physical basis.

In Threads and Spinons Theory:

- The strong force is simply thread confinement at the nuclear scale.
- Protons and neutrons are tightly bound thread loops held together by shared thread connections.
- These threads are highly compressed, preventing protons from repelling due to charge.
- The force weakens beyond the nucleus because thread alignment only supports close- range interactions.

$$F_{\text{strong}} = T_{\text{cocoon}} \cdot SEU$$

where:

- $F_{\text{strong}}$ = Nuclear binding force

- $T_{\text{cocoon}}$ = Tension in the nuclear thread shell

- $SEU$ = Spinon Energy Unit

No need for gluons—thread structure alone explains nuclear stability.

Threads and Spinons Theory
Samir Hanna Safar

Predicts nuclear forces should weaken in unstable isotopes where thread tension is disrupted.

Unifies strong force as an extreme version of thread tension seen in Gravity.

5. The Weak Nuclear Force as Thread Reorganization

In traditional physics:

- The weak force governs radioactive decay and transformations between particles.
- It is mediated by W and Z bosons, but their origin and necessity are unclear.

In Threads and Spinons Theory:

- The weak force is not separate—it is simply the process of thread reconfiguration.
- When a neutron decays, its internal threads realign, releasing a new set of spinons.
- This realignment changes thread connections, allowing particles to transform.

$$F_{\text{weak}} = \frac{T_{\text{nuclear}}}{S_{\text{spinon}}} \cdot E_{\text{decay}}$$

where:

- $F_{\text{weak}}$ = Weak interaction effect
- $T_{\text{nuclear}}$ = Tension inside the decaying nucleus
- $S_{\text{spinon}}$ = Spinon motion inside the particle
- $E_{\text{decay}}$ = Energy released during transformation

No need for W/Z bosons—radioactive decay is a natural process of thread restructuring.

Predicts decay rates based on thread integrity, not quantum randomness.

Unifies the weak force as a thread process, not a separate fundamental interaction.

## 5. The Grand Unified Theory of Forces in Threads and Spinons

| Conventional Physics | Threads and Spinons Explanation |
| --- | --- |
| Gravity is a force or spacetime curvature. | Gravity is thread contraction between masses. |
| Electromagnetism is a field of charge-carried interactions. | Electromagnetism is spinon movement along conductive threads. |
| Strong nuclear force is held together by gluons. | The strong force is thread confinement at nuclear scales. |
| Weak nuclear force causes radioactive decay via W/Z bosons. | The weak force is simply thread realignment and reconfiguration. |

Threads and Spinons Theory
Samir Hanna Safar

- All forces are unified as thread tension and spinon behavior manifestations.

- No need for force particles (gravitons, gluons, W/Z bosons).

- No need for spacetime curvature—Gravity emerges from fundamental thread interactions.

- No need for separate quantum and classical models— everything follows mechanical laws.

Conclusion:

The Ultimate Unification of Physics

The Threads and Spinons Theory unifies Gravity, electromagnetism, and nuclear forces as:

Different manifestations of thread tension and spinon movement. A single framework where actual physical structures determines interactions. A theory that replaces force- carrying particles with deterministic, structured mechanics. A model that works from the quantum scale to galaxies without contradictions. All forces are now connected, replacing fragmented theories with a unified universe vision.

Threads and Spinons Theory
Samir Hanna Safar

**Chapter 39**

## The Origin and Evolution of the Universe Without the Big Bang

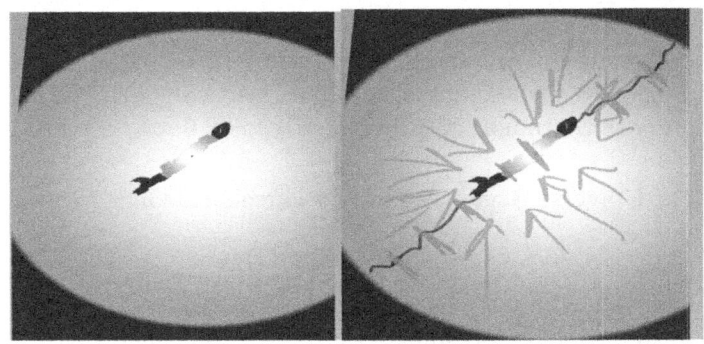

For over a century, the Big Bang Theory has dominated cosmology, proposing that the universe originated from a single point of infinite density and expanded into its current state. However, this theory presents several significant problems:

Singularity Issue: The Big Bang suggests that everything came from "nothing" in an infinitely small point—an unexplained singularity.

Dark Energy & Dark Matter: The theory requires invisible, undetectable forces to explain cosmic expansion and missing mass.

Horizon Problem: Distant regions of the universe have the same temperature, even though Light could not have traveled between them fast enough.

Flatness Problem: The universe appears fine- tuned to be nearly flat, but the reason is unknown.

Expansion without a Cause: The universe's inflationary period suggests rapid growth without a defined mechanism.

The Threads and Spinons Theory eliminates these paradoxes by proposing that the universe did not start with a Big Bang— it has been continuously growing, driven by thread expansion.

1. The Universe Began as a Single Thread and Grew Over Time

The universe did not emerge from a singularity—it began as an infinitely small thread under pressure from an absolute vacuum.

This thread expanded over Time, generating spinons, which created energy and matter.

The expansion is not an explosion but a continuous unwinding of thread structures into larger, more complex forms.

In the beginning:

- The universe contained a single fundamental thread suspended in an absolute vacuum.
- Due to the lack of external forces, this thread expanded under its internal tension.
- As it expanded, it generated spinons, introducing the first energy quanta.
- Over billions of years, this expansion led to the formation of galaxies, stars, and cosmic structures.

$$R_{universe} = k \cdot P$$

where:

- $R_{universe}$ = Universe size over time

- $k$ = Expansion coefficient (based on thread elasticity)

- $P$ = Pressure from the surrounding vacuum

- Explains why the universe is expanding—thread tension continuously unwinds into larger scales.

- Eliminates the need for a singularity—there was no "bang," only gradual growth.

- Accounts for increasing complexity—spinons organize into structured matter over Time.

2. Why Cosmic Expansion is Observed Without Needing Dark Energy

Traditional physics:

- The universe is accelerating, and scientists attribute this to "dark energy," an unknown force.
- However, dark energy has never been detected and remains a mathematical placeholder.

Thread- Based Explanation:

- The universe is expanding because the fundamental thread continues to unwind.
- As the thread stretches, it introduces new spinons, which drive cosmic motion.

Threads and Spinons Theory
Samir Hanna Safar

- Instead of a force like "dark energy," the expansion is a natural property of thread mechanics.

$$a_{expansion} = \frac{T_{thread}}{S_{spinon}} \cdot SEU$$

where:

- $a_{expansion}$ = Rate of cosmic expansion
- $T_{thread}$ = Tension of the fundamental thread
- $S_{spinon}$ = Spinon alignment driving expansion
- $SEU$ = Spinon Energy Unit

- Predicts cosmic expansion without needing an invisible force.

- Explains why expansion accelerates—more spinons = more energy input into large- scale structures.

- Unifies cosmic motion with fundamental thread dynamics.

3. The Cosmic Microwave Background as Residual Thread Tension

Traditional physics:

- The Cosmic Microwave Background (CMB) is seen as the "leftover radiation" from the Big Bang.
- However, CMB radiation is too uniform, raising the horizon problem—distant regions should not have the same temperature.

Thread- Based Explanation:

- The CMB is not the remnant of an explosion—it is the natural background tension of the universal thread system.
- Spinons constantly release small- scale energy fluctuations, creating a consistent low- energy background.
- The universe's large- scale structure influences how thread waves propagate, naturally smoothing out temperature variations.

- No need for the Big Bang—CMB is a continuous thread interaction effect.

- Predicts that CMB temperature fluctuations should match cosmic thread density variations.

- Eliminates the need for inflation—CMB uniformity is a natural thread property.

4. How Galaxies and Stars Formed Without the Big Bang

- Traditional physics:

- Galaxies formed hundreds of millions of years after the Big Bang as minor fluctuations in matter grew due to Gravity.
- However, this model struggles to explain why galaxies are organized and have large- scale structures.

- Thread- Based Explanation:

- Galaxies and stars formed as tensioned thread loops condensed into rotating structures.

- Spinon alignment within thread structures created spiral formations seen in galaxies.

- No need for an initial explosion—galactic structures emerged naturally as threads condensed.

$$M_{\text{galaxy}} = \frac{T_{\text{thread}}}{r} \cdot SEU$$

where:

- $M_{\text{galaxy}}$ = Mass of a galaxy
- $T_{\text{thread}}$ = Thread density in that region
- $r$ = Radius of the galaxy
- $SEU$ = Spinon Energy Unit

- Predicts why galaxies have stable shapes—thread tension organizes them naturally.

- Explains why galaxies have rotational curves without needing "dark matter."

- Unifies cosmic structure with an actual mechanical process, eliminating randomness.

5. The Continuous Growth of the Universe Over Time

- In the traditional Big Bang model, the universe has a finite beginning.

Threads and Spinons Theory
Samir Hanna Safar

- In the Threads and Spinons model, the universe grows continuously as more thread is unwound.

- There is no "beginning"—just a continual expansion of thread structure.

- No sudden singularity—thread tension has always existed, evolving.

- The observable universe is just the portion we can see—the whole thread system may be much more significant.

6. Replacing the Big Bang Model with a Thread Expansion Model

| Big Bang Theory | Threads and Spinons Theory |
| --- | --- |
| Universe started from a singularity. | Universe began as a small fundamental thread and expanded. |
| Expansion is driven by mysterious "dark energy." | Expansion is a natural effect of thread tension. |
| Cosmic Microwave Background is leftover radiation. | CMB is caused by residual thread tension. |
| Galaxies formed randomly after the explosion. | Galaxies emerged naturally as thread loops condensed. |
| Matter originated from a high-energy explosion. | Matter formed through structured spinon energy release. |

- No singularity needed—the universe grew from structured thread expansion.

- No Big Bang explosion—expansion is ongoing and deterministic.

Threads and Spinons Theory
Samir Hanna Safar

- No dark energy—cosmic growth is an intrinsic thread property.

- No fine- tuning problems—large- scale structure follows natural thread patterns.

Conclusion:

The Universe is Expanding, But There Was No Big Bang

The Threads and Spinons Theory redefines cosmic evolution as:

- A gradual, structured expansion of thread tension, not a singular explosion.

- A universe constantly growing without a single moment of creation.

- A natural process where galaxies, stars, and cosmic background radiation emerge from thread interactions.

- A complete rejection of dark matter, energy, and inflation—actual physical mechanics govern everything.

The Big Bang is not needed to explain the universe—only an understanding of thread growth and spinon organization.

Threads and Spinons Theory
Samir Hanna Safar

**Chapter 40**

## The Nature of Time

### Eliminating Time as a Fourth Dimension

Time has been one of the most misunderstood concepts in physics. In classical mechanics, Time is treated as a simple parameter that tracks changes. However, in Einstein's theory of relativity, Time became a fourth dimension supposedly linked to space, forming spacetime.

* The Threads and Spinons Theory eliminates the idea of Time as a fourth dimension, replacing it with fundamental physical changes in thread structures.

1. The Problems with Time as a Fourth Dimension

Traditional physics:

- In Newtonian mechanics, Time is an absolute background that moves forward uniformly.
- In Einstein's relativity, Time is flexible and can be stretched or contracted based on velocity and Gravity.
- Time is treated as an unchangeable parameter in quantum mechanics that governs evolution.

Problems with These Models:

- No physical definition of Time—it is treated as a mathematical tool rather than an actual entity.

- Time dilation suggests time "slows down" at high speeds or near Gravity, but what does this mean?

- Time travel paradoxes arise if Time is a fundamental dimension that can be manipulated.

In the Threads and Spinons Theory, Time is not a separate dimension. Instead, real physical thread tension and spinon movement govern all motion, change, and cause- effect relationships.

2. What is Time in the Threads and Spinons Theory?

- Time is not an actual entity—it is an effect of spinons moving through thread structures.

- All physical processes are just changes in thread tension and spinon organization.

- There is no "flow" of Time—only events happen in a sequence due to thread dynamics.

Threads and Spinons Theory
Samir Hanna Safar

- Time is often described as a "river" flowing forward, but nothing moves except the rearrangement of physical thread connections.

- This replaces "time" with an accurate measure of how fast or slow processes occur.

- There is no separate "time" dimension—just measurable changes in physical thread structures.

3. How This Explains Time Dilation Without a Time Dimension

- In relativity, Time appears to slow down for objects moving at high speeds or near strong Gravity.

- However, this does not mean "time itself" is slowing down—physical processes are happening at different rates.

Thread- Based Explanation:

- At high speeds, spinons take longer to travel through stretched threads, slowing down physical interactions.

- Near massive objects, thread tension increases, affecting how energy moves through the system.

- Clocks tick slower in strong Gravity, not because Time is slowing but because tension physically delays the mechanism.

$$R_{\text{process}} = \frac{T_{\text{thread}}}{S_{\text{spinon}}}$$

where:

- $R_{\text{process}}$ = Rate at which a process occurs

- $T_{\text{thread}}$ = Local thread tension

- $S_{\text{spinon}}$ = Motion speed of spinons in the system


- Explains time dilation without treating Time as a dimension.

- Confirms why clocks slowdown in strong Gravity—thread structure affects spinon movement.

- No paradoxes—just real physical effects on the structure of matter.

4. Why Time Travel is Impossible

- If Time were a fundamental dimension, moving backward in Time should be possible.

- However, no experiment has ever shown evidence of time reversal.

- The Threads and Spinons Theory eliminates time travel.

Why Time Cannot Be Reversed:

- All changes are physical transformations of thread structures.

Threads and Spinons Theory
Samir Hanna Safar

- Once threads rearrange, they cannot be "undone" to their previous state.

- Spinons always move forward along tension paths—they do not "rewind" to past configurations.

Example:

If a cup breaks, the event is recorded in the universe's thread structure. There is no separate "timeline" to go back to—only new thread realignments.

- Time is not a place or a dimension—it is just the sequencing of changes in matter.

- Explains why Time appears to "move forward"—thread structures build upon past states but cannot reverse.

- Unifies cause and effect as purely mechanical interactions, not a flow of Time.

5. Replacing the "Arrow of Time" with Thread Evolution

- Traditional physics treats time as having a forward "arrow" because entropy increases over Time.

- However, entropy is just a statistical effect—it does not explain why the past is inaccessible.

Thread- Based Explanation:

- The "arrow of time" is the natural progression of thread tension releasing energy.

- Once spinons move in a direction, they do not return to a previous state.

- There is no past to return to—only new thread configurations form with each interaction.

- Example:

A burnt match cannot "unburn" because the thread tension structure has permanently changed.

- Time is not flowing—processes occur in a sequence determined by thread mechanics.

## 5. Replacing Spacetime with a Real Physical Framework

| Traditional Physics (Time as a Dimension) | Threads and Spinons Explanation (No Time Dimension) |
|---|---|
| Time is a separate, flowing entity. | Time is just a sequence of physical changes in threads. |
| Time slows down near gravity. | Processes slow down because spinon motion is affected by thread tension. |
| Time can theoretically be reversed. | Thread structures evolve irreversibly—no rewinding possible. |
| Spacetime is curved by mass. | Space is real, but time is just motion along threads. |
| Time is fundamental to the universe. | Time is an illusion created by sequential interactions. |

- No need for "curved spacetime"—only fundamental physical thread interactions.

- Explains why we experience Time in one direction—thread structures only evolve forward.

- Unifies motion, causality, and the universe's evolution under a single physical model.

Threads and Spinons Theory
Samir Hanna Safar

## 7. The Universe Exists Without the Need for Time

- The Big Bang assumes time "began" at a singularity, but this is incorrect.

- If Time does not exist as a dimension, there is no need for a "beginning"—just an ongoing process.

- The Threads and Spinons Theory shows that the universe exists through continuous structural evolution, not time progression.

Conclusion:

- Time is not a fundamental entity—it is just the rate of change in thread structures.

- No need for time travel—thread realignments prevent reversing events.

- No need for a "start" of Time—thread evolution has constantly occurred.

- No need for spacetime curvature—only physical interactions between actual structures.

The concept of Time as a fourth dimension is eliminated—everything is based on actual physical motion, not imaginary timelines.

Threads and Spinons Theory
Samir Hanna Safar

Threads and Spinons Theory
Samir Hanna Safar

**Chapter 41**

## The Rewriting of Black Holes

## The Core Concept

For decades, physicists have treated black holes as one of the universe's most mysterious and paradoxical objects. Stephen Hawking's theory describes them as singularities— infinitely dense points where Gravity becomes so strong that nothing, not even light, can escape.

The Threads and Spinons Theory challenges this idea and replaces black holes with a more realistic concept: The Core.

Cores are not singularities—they are ultra- dense accumulations of threads and spinons, structured and finite, with a definable physical nature.

Threads and Spinons Theory
Samir Hanna Safar

This chapter will explain why black holes, as currently described, are incorrect and how Cores provide a more logical and testable alternative.

1. The Problems with Hawking's Black Hole Model

Hawking's View of Black Holes:

- A black hole forms when a massive star collapses under its Gravity.

- Gravity becomes so strong that it crushes all matter into an infinitely small point (a singularity).

- At the event horizon (the boundary around a black hole), escape velocity exceeds the speed of Light, making it impossible for anything to leave.

- Eventually, black holes evaporate through Hawking radiation—a slow process where particles escape due to quantum fluctuations.

Why This Model Is Flawed:

- Infinite density is impossible—nature does not allow singularities.

- The idea that mass collapses into a single point violates the principles of quantum mechanics.

- If nothing escapes, how does Hawking radiation work? It contradicts the event horizon rule.

- No singularity has ever been directly observed—only gravitational effects have been detected.

- The Threads and Spinons Theory eliminates these paradoxes by redefining black holes as Cores—actual, physical structures with defined properties.

## 2. The Core – A Realistic Alternative to Black Holes

- Cores are ultra- dense thread accumulations, not singularities.

- They are formed when an enormous mass collapses, causing the surrounding threads to contract and pack tightly.

- Instead of "infinitely dense points," Cores have structured layers of compressed threads and spinons.

- Gravity near a Core is extreme but not infinite.

- The Core replaces the black hole singularity with a structured, testable object, removing infinite density and information loss paradoxes.

## What Happens Inside a Core?

- Threads become tightly packed, increasing gravitational tension.

- Spinons slow down, reducing energy emissions (hence why Light cannot escape easily).

- Instead of an event horizon, a Core has a "gravitational density threshold," where threads absorb most incoming Light and matter.

- At a critical point, Cores can release energy in bursts, explaining quasars and gamma- ray bursts.

This means black holes do not "trap" matter forever—they absorb and redistribute energy over Time through thread restructuring.

3. Why do Cores Explain Observations Better Than Black Holes

Current observations of black holes do not match the singularity model—but they align perfectly with the Core model.

Evidence Supporting Cores:

Supermassive black holes (SMBHs) at the center of galaxies:

- If singularities existed, they should "consume" their galaxies over Time.
- Instead, SMBHs seem stable and structured, which is better explained by dense cores balancing gravitational pull and thread tension.

Quasars and High- Energy Emissions from Black Holes:

- Quasars emit powerful radiation from supposed "black holes."
- If nothing escapes a singularity, where does this energy come from?

- The Core model explains this as periodic thread realignments releasing energy in bursts.

Black Hole Mergers and Gravitational Waves:

- LIGO has detected gravitational waves from "black hole collisions."
- Singularities merging should create infinite energy outputs, yet we observe finite waves.
- Cores explain this as structured thread systems merging, creating measurable but finite energy waves.

4. The Core Cycle – Growth, Energy Release, and Rebirth

Instead of being a final dead end, Cores go through a cycle of absorption, compression, and energy redistribution.

Stage 1: Formation

- A dying massive star collapses, causing threads to tighten and spinons to slow down.
- A dense Core forms, replacing the concept of a singularity.

Stage 2: Growth and Matter Absorption

- Matter and energy are pulled in, increasing thread density.
- The Core becomes denser but never infinite.

Stage 3: Energy Redistribution

- Instead of Hawking radiation, Cores release energy when thread structures realign.

- This explains quasars, gamma- ray bursts, and energy jets observed around black holes.

Stage 4: Core Overload and Rebirth

- If too much energy builds up, the Core may explode, releasing stored energy and creating new cosmic structures.
- This could explain why some black holes "disappear" and some galaxies form from intense energy releases.

- Cores are not singularities—they are structured, evolving objects that play a key role in shaping the universe.

5. Predictions and Experimental Tests

- Predictions Based on the Core Model:

- Black holes should have structured layers of density, not singularities.
- Energy should escape in bursts (observed as quasars), not through gradual Hawking radiation.
- Merging black holes should produce measurable but finite gravitational waves, not infinite distortions in spacetime.
- If we study black hole shadows more closely, we should detect thread- like structures in energy emissions, not pure event horizons. The Core model makes testable predictions—something singularities do not.

Conclusion:

Why the Core Model is Superior to Black Hole

| Traditional Black Hole Model | Threads and Spinons Core Model |
|---|---|
| Based on an infinite singularity | Based on ultra-dense but structured threads |
| Requires "curved spacetime" with no physical explanation | Requires real, physical thread tension |
| Predicts information loss, which contradicts physics | Energy is redistributed, not lost |
| Hawking radiation is a mathematical assumption | Quasar emissions and bursts explain energy release |
| Infinite density leads to paradoxes | Density has a real, measurable limit |

- Black holes are not "holes" but dense cores of structured matter and energy built from tightly compressed threads.

- No infinities. No paradoxes. There is no abstract spacetime bending—just real physics with actual structures.

* The Core is the next step in understanding the most extreme objects in the universe.

Threads and Spinons Theory
Samir Hanna Safar

Threads and Spinons Theory
Samir Hanna Safar

**Chapter 42**

**Experimental Proposals for Testing the Threads and
Spinons Theory**

Any scientific theory must be testable and falsifiable to be considered valid. Threads and Spinons Theory introduces a new framework that challenges traditional physics, providing structured explanations for gravity, electromagnetism, quantum mechanics, and cosmology. However, we must design experiments that confirm or refute its claims to move from theory to empirical science.

This chapter outlines key experimental proposals to test the predictions of the Threads and Spinons Theory, providing ways to measure thread tension, spinon transport, and energy interactions.

Testing the Existence of Threads – Measuring Thread Tension in Space

Hypothesis: Space is not empty; it contains an interconnected network of threads that stretch across cosmic distances.

Prediction: If threads exist, then variations in thread tension should be detectable as minute changes in light speed, gravitational effects, or energy flow in space.

Proposed Experiment:

Set up an interferometer-based experiment in deep space (outside of Earth's atmosphere) to measure subtle fluctuations in the speed of light.
Observe if light speed changes when moving in different directions relative to known cosmic structures.
Compare these results to cosmic thread alignments mapped through galaxy surveys.

Expected Outcome:

If light follows thread pathways, its speed should vary slightly depending on thread density.
   This would confirm that light is not traveling through
   space but is transported along structured cosmic threads.

Testing Gravity as a Result of Thread Tension

Hypothesis: Gravity is not curved spacetime but a result of thread contraction.

Prediction: Objects should experience different gravitational effects based on their interaction with cosmic threads, particularly near dense cosmic filaments.

Proposed Experiment:

Use precision accelerometers in different gravitational environments (Earth, the Moon, and deep space) to detect acceleration variations that correlate with expected thread tension differences.

Compare gravitational strength at different altitudes and test whether anomalies exist where thread tension is expected to be lower or higher.
Perform a controlled mass displacement test in orbit to observe whether thread structure influences gravitational pull more than mass alone.

Expected Outcome:
> If gravity is a result of thread contraction, then variations in gravitational effects should correlate with thread density rather than just mass distribution.

Testing the Thread-Based Model of Light Propagation

Hypothesis: Light does not travel through a vacuum but is transported along structured thread pathways.

Prediction: Light behavior should change based on thread alignment, and subtle variations should be measurable in high-precision laser experiments.

Proposed Experiment:

Perform a modified Michelson-Morley experiment with laser beams traveling in different directions relative to known cosmic structures.
Test for minute variations in light speed when traveling parallel vs. perpendicular to cosmic filament alignments.

Analyze high-energy cosmic radiation to check for thread-aligned interference patterns.

Expected Outcome:

> If light follows structured threads, its speed and phase interference patterns should show directional variations, even in a vacuum.

Detecting Spinon Motion and Subatomic Energy Transfer

Hypothesis: Energy is transported through spinon motion along threads rather than as electromagnetic waves in free space.

Prediction: If spinons are accurate, we should be able to manipulate and detect their energy transport separately from traditional electromagnetic radiation.

Proposed Experiment:

Use ultra-low-temperature superconductors to isolate possible spinon-based energy transfer from traditional heat and radiation sources.

Develop a spinon-interaction detector that analyzes energy fluctuations in controlled thread environments.
Experiment with thread-aligned materials to see if energy flows more efficiently in specific directions, similar to superconductors but without electron-based conduction.

Expected Outcome:

If spinons are responsible for energy transport, we should observe non-electromagnetic energy fluctuations consistent with thread structures.

Testing Quantum Entanglement as a Structured Thread Connection

Hypothesis:

Quantum entanglement is not "spooky action at a distance" but a direct interaction between entangled spinons via stretched threads.

Prediction: If entangled particles remain physically connected by threads, then changes to one particle should not be "instantaneous" but detectable as a structured wave propagation effect instead.

Proposed Experiment:

Measure the time delay in quantum entanglement interactions over increasing distances to detect possible wave-like propagation.
Test whether environmental thread tension influences the behavior of entangled particles.
Observe whether manipulating the thread environment affects entanglement coherence.

Expected Outcome:

If entanglement is caused by thread alignment rather than non-local interaction, we should detect structured wave propagation effects instead of instant changes.

Threads and Spinons Theory
Samir Hanna Safar

Re-Evaluating Cosmic Redshift as a Function of Thread Expansion

Hypothesis: Redshift is not due to Doppler expansion of space but is an effect of thread elongation over time.

Prediction: If redshift is caused by thread structure, then redshift values should correlate with cosmic web density, not just distance.

Proposed Experiment:

Analyze galaxy redshift maps and compare them to the density of cosmic filaments.
Check if redshift varies based on local thread density rather than simply increasing with distance.
Develop an experiment that measures thread elongation at more minor scales to test its effect on light properties.

Expected Outcome:
> If redshift is a thread-based effect, then the standard model's assumption of an expanding universe may need to be revised.

Testing Artificial Gravity via Thread Manipulation

Hypothesis: If gravity is caused by thread contraction, then controlled thread realignment should be able to generate artificial gravity. Prediction: If we can manipulate thread structures, we should be able to create localized gravitational fields without mass-based attraction.

Proposed Experiment:
Develop a high-intensity thread-tension generator to induce thread contractions artificially.

Threads and Spinons Theory
Samir Hanna Safar

Test whether an object in a controlled environment experiences artificial gravitational effects when thread tension is altered.
Compare results with expected gravitational effects under standard Newtonian physics.

Expected Outcome:

If thread tension governs gravity, artificial gravity should be achievable through controlled thread contractions.

Conclusion:

The Path to Empirical Validation

| Traditional Physics Tests | Threads and Spinons Experimental Approach |
| --- | --- |
| Tests assume light moves freely in a vacuum | Tests measure light's structured transport along threads |
| Gravity experiments assume mass curves spacetime | Tests measure gravity as a function of thread contraction |
| Quantum entanglement assumes instantaneous interaction | Tests analyze structured wave propagation effects |
| Cosmic redshift assumes space expansion | Tests check for redshift correlations with thread density |

The Threads and Spinons Theory is not just a theoretical framework but a testable model that can be experimentally validated.
These experiments provide a roadmap for future scientific breakthroughs, challenging mainstream assumptions and offering a new way to understand the universe.

Physics future is structured, testable, and observable science—this theory provides the foundation for that shift.


Threads and Spinons Theory
Samir Hanna Safar

.

Threads and Spinons Theory
Samir Hanna Safar

# VI

# Closing Chapter

Threads and Spinons Theory
Samir Hanna Safar

**Closing Chapter**

**A Theory in Progress**

**The Path to Refinement and Justification**

No theory is perfect. Without rigorous testing, refinement, and validation, no framework can claim to be the absolute truth about the universe. The Threads and Spinons Theory is no exception. It is a bold attempt to unify Gravity, electromagnetism, quantum mechanics, and Consciousness into one fundamental structure—but it is just the beginning.

This theory provides a solid foundation but must be continuously refined, tested, and challenged.

1. Why No Theory is Ever Final

- Science is an evolving process, not a static collection of facts.

- Newton's Gravity worked for centuries until Einstein showed it was incomplete.

Threads and Spinons Theory
Samir Hanna Safar

- Einstein's Relativity is brilliant but fails to integrate quantum mechanics.

- Quantum mechanics is mathematically robust but lacks a physical explanation.

- Every grand theory begins with a first step—a best attempt that future scientists will refine.

The Threads and Spinons Theory provides a structured framework for understanding:

- Gravity is a result of thread tension instead of spacetime warping.
- Light as structured spinon motion, eliminating wave-particle duality paradoxes.
- Quantum mechanics is defined as deterministic thread interactions rather than probabilistic waves.
- The universe is an ever- growing expanding thread network, not a Big Bang explosion.

However, this is not the final version of the theory—it is the first strong model that will evolve with further research and experimentation.

2. What Still Needs Refinement?

- Thread Structure at the Smallest Scale:

- How thin are the threads?
- Can we measure their tension directly?
- Are there substructures within spinons?

Experimental Confirmation:

- Can we detect variations in light speed caused by thread tension?
- Can we confirm gravity propagation as a function of thread elasticity?
- Can we manipulate spinon stacks to create new energy systems?

Mathematical Formalization:

- While we have provided equations, further mathematical refinement is needed to match observational data more precisely.
- More work is needed to link these equations directly to existing physics models, which would facilitate their adoption by the scientific community.

These are exciting challenges—not roadblocks. The most significant discoveries in history started with incomplete theories that were improved over time.

3. Why This is a Strong Starting Point for a Unified Theory

- It eliminates paradoxes that have haunted modern physics for decades (e.g., wave- particle duality, dark Matter, quantum randomness).

- It provides physical explanations instead of just mathematical descriptions.

- It connects major forces into a single framework rather than treating them as separate entities.

- It introduces testable predictions, which allows future researchers to confirm or refine the model.

Threads and Spinons Theory
Samir Hanna Safar

Even if parts of this theory require revision, its core ideas—threads, spinons, and structured energy—offer a promising pathway to unifying physics.

4. A Unified Universe – The Ultimate Goal

- Science and philosophy have long searched for a unified explanation of existence.

- This theory brings us closer to that goal by offering a model where everything—energy, Matter, Light, Gravity—is interconnected.

- Instead of fragmented forces and separate particles, we now have a system based on structured, physical connections.

- This model does not just unify physics—it unifies our understanding of creation itself.

- Threads and spinons are not just mechanical parts of the universe; they represent the very fabric of existence.

5. The Call to Action – A New Scientific Revolution

- This theory cannot advance without testing, discussion, and refinement.

- Scientists, engineers, and independent thinkers must come together to test their predictions.

- New experiments must be designed to detect thread tension, spinon motion, and structured energy patterns.

This book is not the end—it is an invitation.

Threads and Spinons Theory
Samir Hanna Safar

- The future of physics is in our hands.

- Let us question everything.

- Let us challenge existing models.

- Let us refine and justify this theory with actual experimental data.

- And above all, let us continue searching for the truth about our universe.

Final Thought:

This is Just the Beginning

The Threads and Spinons Theory is a bold new way of looking at the universe. It challenges conventional Science while building upon its successes. It provides a structured foundation for a true Theory of Everything. But it is not perfect. Moreover, that is okay. Science is about continuous discovery, refinement, and evolution. What matters is that we have taken the first step toward understanding the universe in a way never done before. With time, testing, and collaboration, this theory will evolve even more.

his book is the spark—the fire of discovery is now in your hands.

**The journey continues ...**

# Reference

This section references the key scientific concepts that the Threads and Spinons Theory builds upon, challenges, or refines. It includes foundational works in physics, quantum mechanics, relativity, cosmology, nuclear physics, and quantum computing. These references serve as a comparative framework, demonstrating how this theory aligns with or diverges from mainstream scientific understanding.

1. Classical Physics and Gravity

- Newton, I. (1687). *Philosophiæ Naturalis Principia Mathematica*. London: Royal Society.

    • Foundation of classical mechanics and Newtonian gravity.

- Einstein, A. (1915). *Die Grundlage der allgemeinen Relativitätstheorie* (The Foundation of General Relativity). *Annalen der Physik, 354*(7), 769- 822.

    • Proposed the curvature of spacetime as the mechanism of gravity.

- Misner, C. W., Thorne, K. S., & Wheeler, J. A. (1973). *Gravitation*. San Francisco: W. H. Freeman.

    • Detailed analysis of General Relativity and spacetime curvature.

- Weinberg, S. (1972). *Gravitation and Cosmology: Principles and Applications of the General Theory of Relativity.* Wiley.

  • Overview of modern gravitational theories and cosmology.

- Relevance to This Theory:
- Newton's model explains forces but lacks a mechanism.
- Einstein describes gravity as curvature but does not explain what space is made of.
- Threads and Spinons replace these models with a physical structure—gravity emerges from thread tension.

2. Quantum Mechanics and Electromagnetism

- Planck, M. (1900). *On the Law of Distribution of Energy in the Normal Spectrum. Annalen der Physik, 4*(553), 1- 10.

  • Introduced quantization of energy.

- Bohr, N. (1913). *On the Constitution of Atoms and Molecules. Philosophical Magazine, 26*(1), 1- 25.

  • Introduced quantum jumps for atomic electrons.

- Heisenberg, W. (1927). *Über den anschaulichen Inhalt der quantentheoretischen Kinematik und Mechanik. Zeitschrift für Physik, 43*, 172- 198.

  • Formulated the Uncertainty Principle.

- Schrödinger, E. (1926). *Quantisierung als Eigenwertproblem. Annalen der Physik, 79*(4), 361- 376.

- Developed wave mechanics and quantum states.

- Maxwell, J. C. (1865). *A Dynamical Theory of the Electromagnetic Field. Philosophical Transactions of the Royal Society of London, 155*, 459- 512.

  - Established Maxwell's Equations for electromagnetism.

- Relevance to This Theory:
- Quantum mechanics relies on probability and wave- particle duality—this theory provides a structured alternative.
- Light is not a probability wave but structured spinon motion along threads.
- Maxwell's equations describe electromagnetism but do not explain its medium—this theory replaces "fields" with structured threads.

3. The Big Bang and Cosmology

- Lemaître, G. (1927). *Un Univers homogène de masse constante et de rayon croissant rendant compte de la vitesse radiale des nébuleuses extra- galactiques. Annales de la Société Scientifique de Bruxelles, 47*, 49- 59.

  - Proposed the expanding universe and early version of the Big Bang model.

- Gamow, G. (1948). *The Origin of Chemical Elements. Physical Review, 74*(4), 505- 526.

  - Predicted the cosmic microwave background radiation.

- Penzias, A. A., & Wilson, R. W. (1965). *A Measurement of Excess Antenna Temperature at 4080 Mc/s. The Astrophysical Journal, 142*, 419- 421.

- Discovery of CMB radiation, considered proof of the Big Bang.

- Relevance to This Theory:
- The Big Bang assumes a singularity—this theory replaces it with continuous thread expansion.
- The CMB is not "leftover heat" but a structured thread-network signature.

4. Black Holes and Hawking Radiation

- Schwarzschild, K. (1916). *Über das Gravitationsfeld eines Massenpunktes nach der Einstein'schen Theorie. Sitzungsberichte der Königlich Preussischen Akademie der Wissenschaften, 7*, 189- 196.

- First exact black hole solution in General Relativity.

- Hawking, S. W. (1974). *Black Hole Explosions? Nature, 248*(5443), 30- 31.

- Proposed Hawking Radiation, where black holes slowly evaporate.

- Relevance to This Theory:
- Black holes are not singularities—they are dense cores of structured thread accumulations.
- Hawking radiation assumes a loss of information—this model explains energy redistribution through thread realignment.

5. Quantum Computing and Subatomic Information Storage

- Feynman, R. P. (1982). *Simulating Physics with Computers. International Journal of Theoretical Physics, 21*(6- 7), 467-488.

- Proposed quantum computing as a simulation tool for quantum mechanics.

- Deutsch, D. (1985). *Quantum Theory, the Church- Turing Principle, and the Universal Quantum Computer. Proceedings of the Royal Society A: Mathematical, Physical and Engineering Sciences, 400*(1818), 97- 117.

- Developed the theory behind quantum computing algorithms.

- Relevance to This Theory:
- Quantum computing assumes probability- based operations—this model replaces it with structured spinon transport.
- Quantum memory is not based on superposition but on structured energy storage in thread networks.

6. Experimental Physics and Future Testing

- LIGO Scientific Collaboration. (2016). *Observation of Gravitational Waves from a Binary Black Hole Merger. Physical Review Letters, 116*(6), 061102.

First detection of gravitational waves, proving massive energy shifts.

- Riess, A. G., et al. (1998). *Observational Evidence from Supernovae for an Accelerating Universe and a Cosmological Constant. The Astronomical Journal, 116*(3), 1009- 1038.

- Evidence of cosmic acceleration, leading to dark energy models.

- Relevance to This Theory:
- LIGO waves suggest structured Core mergers, not singularities.
- Redefining gravity through thread tension could produce new experiments to test these ideas.

**Final Notes:**

These references provide the scientific foundation that the Threads and Spinons Theory builds upon, refines, or replaces. While much of this theory is new, it is crucial to show how it challenges, corrects, and improves upon existing research.

Made in the USA
Las Vegas, NV
20 March 2025

19870501R00223